GREEN ICT & ENERGY: FROM SMART TO WISE STRATEGIES

Sustainable Energy Developments

Series Editor

Jochen Bundschuh
University of Southern Queensland (USQ), Toowoomba, Australia
Royal Institute of Technology (KTH), Stockholm, Sweden

ISSN: 2164-0645

Volume 5

Green ICT & Energy

From smart to wise strategies

Jaco Appelman

*Assistant Professor Systems Engineering, Delft University of Technology
and Sustainability Innovator at FastFact.nl, Delft, The Netherlands*

Anwar Osseyran

*CEO SURFsara, the Dutch national HPC Center and Chairman of the Green IT Consortium
Amsterdam, The Netherlands*

Martijn Warnier

*Assistant Professor Systems Engineering, Delft University of Technology,
Delft, The Netherlands*

CRC Press
Taylor & Francis Group
Boca Raton London New York Leiden

CRC Press is an imprint of the
Taylor & Francis Group, an **informa** business

A BALKEMA BOOK

Short note on front cover:
The energy vortex resembles, if you want to see it, DNA and all life is essentially a mixture of flows of energy and stored energy. We hope that new ideas about sustainability will bring more life to the ICT sector so that it can flourish and grow and contribute to new green prosperous world.

CRC Press/Balkema is an imprint of the Taylor & Francis Group, an informa business

© 2014 Taylor & Francis Group, London, UK

Typeset by MPS Limited, Chennai, India
Printed and bound in The Netherlands by PrintSupport4U, Meppel

Library of Congress Cataloging-in-Publication Data

Green ICT & energy : from smart to wise strategies / [edited by] Jaco Appelman,
    Anwar Osseyran & Martijn Warnier.
        pages cm. — (Sustainable energy developments, ISSN 2164-0645; volume 5)
    Includes bibliographical references and index.
    ISBN 978-0-415-62096-3 (hardback : alk. paper)   1. Information technology—
Environmental aspects.   2. Computer science—Environmental aspects.   3. Computer
industry—Environmental aspects.   4. Green technology.   I. Appelman, Jacob Herre.
II. Osseyran, A.   III. Warnier, Martijn.
    QA76.9.E58G74   2014
    004—dc23
                                                                          2013027376

Published by: CRC Press/Balkema
                P.O. Box 11320, 2301 EH, Leiden, The Netherlands
                e-mail: Pub.NL@taylorandfrancis.com
                www.crcpress.com – www.taylorandfrancis.com

ISBN: 978 0 415 62096 3 (Hardback)
ISBN: 978 0 203 12023 1 (e-book PDF)

About the book series

Renewable energy sources and sustainable policies, including the promotion of energy efficiency and energy conservation, offer substantial long-term benefits to industrialized, developing and transitional countries. They provide access to clean and domestically available energy and lead to a decreased dependence on fossil fuel imports, and a reduction in greenhouse gas emissions.

Replacing fossil fuels with renewable resources affords a solution to the increased scarcity and price of fossil fuels. Additionally it helps to reduce anthropogenic emission of greenhouse gases and their impacts on climate change. In the energy sector, fossil fuels can be replaced by renewable energy sources. In the chemistry sector, petroleum chemistry can be replaced by sustainable or green chemistry. In agriculture, sustainable methods can be used that enable soils to act as carbon dioxide sinks. In the construction sector, sustainable building practice and green construction can be used, replacing for example steel-enforced concrete by textile-reinforced concrete. Research and development and capital investments in all these sectors will not only contribute to climate protection but will also stimulate economic growth and create millions of new jobs.

This book series will serve as a multi-disciplinary resource. It links the use of renewable energy and renewable raw materials, such as sustainably grown plants, with the needs of human society. The series addresses the rapidly growing worldwide interest in sustainable solutions. These solutions foster development and economic growth while providing a secure supply of energy. They make society less dependent on petroleum by substituting alternative compounds for fossil-fuel-based goods. All these contribute to minimize our impacts on climate change. The series covers all fields of renewable energy sources and materials. It addresses possible applications not only from a technical point of view, but also from economic, financial, social and political viewpoints. Legislative and regulatory aspects, key issues for implementing sustainable measures, are of particular interest.

This book series aims to become a state-of-the-art resource for a broad group of readers including a diversity of stakeholders and professionals. Readers will include members of governmental and non-governmental organizations, international funding agencies, universities, public energy institutions, the renewable industry sector, the green chemistry sector, organic farmers and farming industry, public health and other relevant institutions, and the broader public. It is designed to increase awareness and understanding of renewable energy sources and the use of sustainable materials. It also aims to accelerate their development and deployment worldwide, bringing their use into the mainstream over the next few decades while systematically replacing fossil and nuclear fuels.

The objective of this book series is to focus on practical solutions in the implementation of sustainable energy and climate protection projects. Not moving forward with these efforts could have serious social and economic impacts. This book series will help to consolidate international findings on sustainable solutions. It includes books authored and edited by world-renowned scientists and engineers and by leading authorities in economics and politics. It will provide a valuable reference work to help surmount our existing global challenges.

Jochen Bundschuh
(Series Editor)

Editorial board

Morgan Bazilian   Deputy Director, Institute for Strategic Energy Analysis (JISEA), National Renewable Energy Lab (NREL), Golden, CO, USA, morgan.bazilian@nrel.gov

Robert K. Dixon   Leader, Climate and Chemicals, The Global Environment Facility, The World Bank Group, Washington, DC, USA, rdixon1@thegef.org

Maria da Graça Carvalho   Member of the European Parliament, Brussels & Professor at Instituto Superior Técnico, Technical University of Lisbon, Portugal, maria.carvalho@ist.utl.pt, mariadagraca.carvalho@europarl.europa.eu

Rainer Hinrichs-Rahlwes   President of the European Renewable Energy Council (EREC), President of the European Renewable Energies Federation (EREF), Brussels, Belgium; Board Member of the German Renewable Energy Federation (BEE), Berlin, Germany, rainer.hinrichs@bee-ev.de

Eric Martinot   Senior Research Director, Institute for Sustainable Energy Policies (ISEP), Nakano, Tokyo & Tsinghua University, Tsinghua-BP Clean Energy Research and Education Center, Beijing, China, martinot@isep.or.jp, martinot@tsinghua.edu.cn

Veena Joshi   Senior Advisor-Energy, Section Climate Change and Development, Embassy of Switzerland, New Delhi, India, veena.joshi@sdc.net

Christine Milne   Leader of the Australian Greens Party, Senator for Tasmania, Parliament House, Canberra, ACT & Hobart, TAS, Australia

ADVISORY EDITORIAL BOARD

ALGERIA

Hacene Mahmoudi (renewable energy for desalination and water treatment), Faculty of Sciences, Hassiba Ben Bouali University, Chlef

ARGENTINA

Marta Irene Litter (advanced oxidation technologies, heterogeneous photocatalysis), Gerencia Química, Comisión Nacional de Energía Atómica, San Martín, Prov. de Buenos Aires, Argentina & Consejo Nacional de Investigaciones Científicas y Técnicas, Buenos Aires, Argentina & Instituto de Investigación e Ingeniería Ambiental, Universidad de General San Martín, Prov. de Buenos Aires

AUSTRALIA

Thomas Banhazi (biological agriculture; sustainable farming, agriculture sustainable energy solutions), National Centre of Engineering in Agriculture, University of Southern Queensland, Toowoomba, QLD

Table of contents

List of contributors

Dayna Baumeister has a background in biology, a devotion to applied natural history, and a passion for sharing the wonders of nature with others. She has worked in the field of biomimicry with Janine Benyus since 1998 as a business catalyst, educator, researcher, and design consultant. Together they founded the Biomimicry Guild, The Biomimicry Institute, and Biomimicry3.8, collectively fertilizing the movement of biomimicry as an innovative practice and philosophy to meet the world's sustainability challenges. Dayna designed and teaches the world's first Biomimicry Professional Certification Program and compiled over a decade of experience in biomimicry into the *Biomimicry Resource Handbook: A Seed Bank of Knowledge and Best Practices* (2013).

Erik Beulen holds the Global Sourcing chair at Tilburg University, The Netherlands and is also employed by KPMG.

Freek Bomhof completed his M.Sc. Electrical Engineering in 1990. During his work on optimization of business processes through innovative ICT's, he became interested in the relation between technology and human behavior in organizations, especially when considering the sustainability effects of that behavior. With his background in pattern recognition and project management, he tries to quantify this behavior and its effects in terms of indicators that can be managed from a sustainability viewpoint.

Frances Brazier is a full professor in Engineering Systems Foundations, within the Faculty of Policy and Management, at the Delft University of Technology, as of September 2009, prior to which she chaired the Intelligent Interactive Distributed Systems Group at the VU University, Amsterdam. Her group's research focuses primarily on the design and (self) management of large-scale distributed autonomous (adaptive) systems in dynamic environments.

Kassidy Clark received his M.Sc. in Computer Science at the VU University, Amsterdam, the Netherlands. He is currently pursuing a Ph.D. in Computer Science at the Delft University of Technology, The Netherlands. His research focusses on service negotiation in large scale, distributed systems, including the Cloud and the Smart Energy Grid.

Tineke M. Egyedi is founding director of the Delft Institute for Research on Standardization (DIRoS) and senior researcher at the Delft University of Technology. She initiated the game "setting standards" (developed with United Knowledge), which has been/is used to teach students (several European universities), US policy makers (NIST), and Chinese standardizers (NEN). She is board member of the European Academy for Standardization (EURAS).

Johanne Punte Kalsheim is a graduate student from Delft University of Technology, The Netherlands and is employed by Ernst & Young in Norway.

Heide Lukosch is an Assistant Professor at Delft University of Technology. Her research questions focus on social processes within virtual training environments and simulation games. She is exploring how simulation games can help increase the situational awareness of teams in safety-critical domains. This includes also the use of concepts of (formal and informal) learning, microtraining and microgaming, game-based learning, and virtual worlds.

Karel Mulder graduated in Philosophy of Science, Technology & Society, at the University of Twente and completed a Ph.D. at University of Groningen, Faculty of Management studies. Since 1992 Karel Mulder has worked at TU Delft as Lecturer "Technology Assessment" at faculty Technology Policy &

Management, TU Delft and became project leader of the Project Sustainable Development in Engineering Education at TU Delft. Since 1999 he has been an Associate Professor and Head of Technology Dynamics & Sustainable Development unit.

Sachiko Muto is a Ph.D. candidate at Delft University of Technology and currently a visiting researcher at UC Berkeley, Center for Science, Technology, Medicine & Society. Her work focuses on government involvement in the making of interoperability standards to achieve public policy goals, e.g. for Smart Grid and eHealth.

Reinier de Nooij has a Ph.D. in environmental science. He first specialized in biology and biodiversity protection and later also in the Framework for Strategic Sustainable Development (also known as The Natural Step). Since 2008, he works as an independent trainer and advisor in sustainability of spatial projects.

Mariette Overschie is a researcher and lecturer at the Delft University of Technology, Faculty of Technology, Policy and Management, in the field of technology and sustainable innovation. She holds a Master in Science Degree in Industrial Design Engineering, Design for Sustainability. At the moment she is preparing her Ph.D. thesis at TU Delft on the topic of microtraining to support sustainability in organizations.

Dirk Jan Peet graduated in Chemical Engineering, specializing in bio-process technology. He has developed and taught courses in Technology in Sustainable development at the faculties of Electrical Engineering, Technical Mathematics, Applied Earth Sciences, and Chemical Engineering. Dirk Jan presented a first version of a part of this paper in 2005. Unfortunately, Dirk Jan suffered from a chronic disease. He passed away on 22 of November 2008.

Henk Plessius graduated at the University of Twente and has worked in ICT-education ever since. His experience ranges from lecturer and curriculum developer to consultant and manager. Apart from Green IT, he is interested in problems regarding the interface between organization and information, as well as in modeling and architecture.

Pieter de Vries is Assistant Professor at the TUD coming from the private sector where he is still engaged. He has an interdisciplinary background in Human Geography, Mass Communication, Instructional Technology, and Organizational Learning and Technology. His research focus is on technology enhanced learning, learning value management, informal, and autonomous learning and comprises industries and higher education. He chairs the workgroup "Technologies for Learning" of the European Society for Engineering Education.

Foreword

This book is the result of our personal journeys in the ICT-sector and a great deal of literature but most of all the result of our shared conviction that ICT can and will need to play an important role in the 'biological age' we are moving into. Cyclical glocalized interdependent and resilient or robust production- and service systems will need to be designed, constructed, governed, maintained and evolve. Next-generation systems and infrastructures will need to be adaptable and provide stability. Quite a challenge because at face value the concepts of adaptability and stability seem to be opposed. Most natural systems however 'know' very well how to provide stability while leaving room or retaining agency to adapt to changing circumstances.

In this sense the ICT-sector is at the beginning of things. It is still a sector that creates waste and destroys life while making and discarding products. Our current linear economies will become cyclical and bio-based and thus need to deal with limited resources. The ICT sector is learning how the production of the tools we work with and the content with which we fill and fuel these tools can contribute to economic and ecological wealth and helps or at least does not impede the development of socially just, pleasant and elegant societies.

The reader, engaged in sustainability will find this book probably a bit too conservative and positive, while readers from industry will sometimes feel that that they are being pushed and presented with really weird and far-out ideas that make no economic business sense.

The book builds along these lines, the first introductory chapters sketch the need and magnitude of the challenge we will face the coming decades. And gradually more and more ideas and solutions are given and discussed. Different fields like education will need to contribute to a transformation of a sector and the chapter before the conclusions highlights an approach that is well compatible with blue or green economy initiatives. Using ICT to help green other sectors is also part of this volume, there is however a great deal of literature on the implicit and more visible benefits of employing ICT. While literature on greening the ICT-sector itself seem conspicuously absent. After the introduction in the first 3 chapters we explore this subject in the last two chapters.

It clearly shows how much we can still learn, improve and innovate. The challenge is not only to research, design and build more powerful computing devices such fractal- and quantum computers. It is of equal importance to design in such a way that products and systems become beneficial when they operate not only when the technology is used for 'good' purposes.

This vision on ICT delivers an extremely exciting journey of discovery where product design and use of products is concerned. We are at the brink of a new era where the whole planet can participate in our global society.

We hope that this book motivates many people to embark on this journey to green ICT and create new competitive advantages that benefit a healthy planet and where people feel that they can contribute substantially for that to become reality we will need to work in and through an ICT-supported world.

Anwar Osseyran
Martijn Warnier
Jaco Appelman
(editors)

Preface by Prof. W. Vermeend

While reading this book it struck me again how pervasive and important the ICT-industry is. An industry that underpins and connects literally the whole world. The ICT industry strongly supports and pursues a more efficiently functioning industry. It is smart to focus on efficiency and it is needed.

It is wise if we can come up with designs and solutions that aim to: "... generate value in more than one domain and once installed for a long period of time. The main thing is that you go further than reduction of energy and resource-use, the rationale underlying all efficiency approaches." (Ch3 of this publication).

This book is a first exploration of that theme. In sustainability you will find the distinction between eco-efficiency and eco-effectiveness. When adopting an eco-efficient approach the focus is on re-use, reduction and recycling, a 'do-more with less' strategy. Such strategies can easily rebound in a net increase in material and energy use if not accompanied by Eco-effective approaches. This book gives input from different scientific fields to show that incorporation of eco-effective goals is possible and do-able.

It is not a 'do-it-yourself' book, a handbook or manual for change. The focus is to acquaint the reader with the subject matter of sustainability and ICT, because a focus on this subject will have positive consequences on energy used. This book shows how through education, regulation, ICT-support, standardization and use of technology steps can be made and a strategy forged. The authors conclude that we are still at the beginning of eco-effective strategies and solutions, but acknowledging this point is the first step toward more eco-effective or wise strategies.

Prof. Willem Vermeend
chairman of the topsector ICT
fiscal and financial specialist
served as minister of state and
other political functions

About the editors

Jaco Appelman is a part-time senior researcher and assistant professor at the Delft University of Technology. He specializes in process support for participatory systems design and is involved in a KIC-Climate project to develop a waterfootprint based strategy tool for corporations. Main area's of application are: Safety, Sustainability and Health. Currently involved in a Green Deal Biomimicry where the focus is on stimulating Water Based Chemistry as part of the transition to a bio-based or circular economy and advisor to the Biomimicry InstituteNL and member of the Dutch Footprint Network and member Member Program Committee of the IADIS International Conference on Sustainability, Technology and Education (STE 2012) Jaco is committed, creative & integrating, enthusiastic and pragmatic with a background in business and social sciences. As an entrepreneur Jaco combines facilitation, gaming, business, coaching and personal development methods to provide process support and programme management for bio-based (system-) innovations and implementation trajectories.

Specialties: sustainability, bio-based innovations, e-support for participation, Green(ing)-IT, programme management, (on-line) learning.

Anwar Osseyran is since 2001 the CEO of SURFsara (formerly SARA), The Dutch national HPC Center. He has more than 30 years of multidisciplinary management experience in various areas of ICT including manufacturing, information management, health informatics, high performance computing, industrial automation and computer hardware and software. He initiated in 2009 the Green IT Consortium Amsterdam, a public-private partnership aimed at reducing carbon emissions within datacenters and deploying ICT to stimulate a green economy in the Amsterdam region.

Anwar Osseyran has hands-on experience in working within multinational organizations and has lived and worked in Lebanon, France, The Netherlands, UK, Japan and the USA. He is the chairman of the Advisory board of the Swiss National Supercomputing Centre, member of the of the Scientific Advisory Council of the Cyprus Institute for Supercomputing, member of the Executive Board of Nederland ICT (the association of the Dutch ICT sector), member of the Amsterdam Climate Council, member of the board of Almere Data Capital and member of the Executive board of the Alan Turing Institute Almere. Anwar Osseyran is co-author of the book "Broadband in a world of Glass", ISBN 903952449 and of the book Duurzame ICT, ISBN 9789012582285.

Specialties: cloud computing, green IT, supercomputing, broadband, industrial automation, health informatics, testing of electronics and simulation of electromagnetic fields.

Martijn Warnier is a senior researcher and assistant professor at the Delft University of Technology. He has published more than 70 scientific papers ranging from subjects such as mathematics and computer security to philosophy and multi-agent systems. Currently his main research focus is on (self-)management for large scale socio-technical systems: systems that combine people and organizations with technical infrastructures such as the internet, road networks or the electricity grid. Efficient use of resources has always been a main concern of Martijn's research and by trying to enhance the efficiency of the electricity grid his research became part of the green or environmental computing field.

Specialties: efficiency, (self-)management for large socio-technical systems, multi-agent systems.

CHAPTER 1

Editorial introduction: A strategic view on changing an industry

Jaco Appelman, Anwar Osseyran & Martijn Warnier

1.1   INTRODUCTION

The past 40 years witnessed the emergence of a totally new industry and services sector: the Information and Communication Technology sector, commonly referred to as the ICT sector. During those decades, government, industry, and consumers have become dependent on ICT to coordinate stocks and flows of people, goods, materials, and services around the world. This growth has been healthy from an economic perspective, creating and replacing many jobs that were lost in other parts of industry. The growth was also motivated by the efficiency improvement generated by ICT. But the other side of the coin is the fact that efficiency improvements lead to higher consumption and further increases in energy and material use. Growth cannot be sustained if we remain producing in a cradle-to-grave fashion.

We will sketch why it is prudent for the ICT sector to try to become a more "green" sector in the background section and come to the conclusion that the current efforts of the ICT industry leaders are primarily aimed at becoming more efficient. This is a laudable goal but insufficient in the long run because increases in population and dwindling resource bases force us to be more radically innovative. If we do this right, the payoffs economically, ecologically, and socially of *green IT* are potentially enormous. But what exactly is green IT? In an overview, San Murugesan defines green IT as "the study and practice of designing, manufacturing, using, and disposing of computers, servers, and associated subsystems—such as monitors, printers, storage devices, and networking and communications systems—*efficiently and effectively* [emphasis added] with minimal or no impact on the environment." This definition leaves out an important aspect of ICT more appropriately captured by Gartner in 2007[1] as ". . . the optimal use of ICT for managing the environmental sustainability of operations and the supply chain, as well as that of its products, services and resources, throughout their life cycles." Optimal use means not only greening the ICT infrastructure itself (manufacturing, using, and disposing) but also deploying ICT for greening applications across the whole life cycle. At this point, it is good to observe that Murugesan also employs the efficiency and effectiveness dichotomy. The way he frames effectiveness is however quite generic and we need to further specify what effectiveness means.

When we consider an effectiveness strategy the eco-prefix is imperative because it indicates that, to be successful in the long run, you will need to find resources that are indefinitely renewable, stay within ecological bounds. This means that we can use all biologically produced materials and a very limited amount of mineral resources and metals because the regeneration cycles of these substances are geologically defined and geological processes do have impressive time scales. The effect you are trying to reach with such a strategy is that you create more life unlike the current economic strategies that destroy life. And for those who still not got it, we are living beings. We can create more life and more economic growth and development if we aim to develop products and production platforms that create value in at least three dimensions and not just the financial one. That is to say beneficial to the natural environment, inclusive at the social level, and financially profitable in the economic realm.

---

[1]Gartner RAS Core Research Note G00153703.

In the following section, we start with an impression of the economic performance of the industry and continue with energy use and e-waste to sketch the ecological and social performance of the sector. Energy use and e-waste are discussed because, taken together, these are good indicators of the relative ecological footprint of the production of most goods (Huijbregts *et al.*, 2008; 2010). Then we report on the initiatives undertaken by the ICT sector itself to decrease its impact on the environment while remaining profitable. We unite the sustainability subsection and the ICT subsection in the subsequent conclusions where we outline a strategy, applicable at all levels and therefore necessarily rather abstract, that can make the ICT industry "last forever".

1.2   STATE OF THE ART: THE ICT SECTOR

Economic growth
From an economic point of view, the ICT sector is performing above average and the trend is, under similar circumstances, that this growth pattern will continue. Initially the ICT sector has had a relatively green image, due to the promises it potentially holds to support the greening of other sectors. The ICT industry has experienced impressive growth figures over the past 30 years. The sector contributed 16% of global GDP growth from 2002 to 2007 and the sector itself has increased its share of GDP worldwide from 5.8 to 7.3% and it is expected to jump further to 8.7% of GDP growth worldwide in 2020. In low-income countries, an average of 10 more mobile phone users per 100 people was found to stimulate a per capita GDP growth of 0.59% (The Climate Group, 2008). The report of the Climate Group further suggests that a third of the economic growth in the Organization for Economic Cooperation and Development (OECD) countries between 1970 and 1990 was due to access to fixed-line telecom networks alone, which lowered transaction costs and helped firms to access new markets.

Energy efficiency in manufacturing
The energy consumed by a manufacturing process is a major direct measure of its impact on the environment. The energy consumed usually translates to the amount of energy that has been produced from fossil-fuel-fired plants or captive generators. The energy consumed thus has a strong link with the amount of fossil fuels consumed and contributes therefore to the depletion and degradation of the environment (soil, air, and water).

A typical desktop personal computer (PC) with a 17″ flat panel LCD monitor requires about 100 Watts. Not much? Left on 24/7 for one year, the system will consume a whopping 876 kWh electricity. That is enough to release approximately 360 kg of carbon dioxide into the atmosphere—the equivalent of driving about 1300 kilometres in an average car (NASSCOM, 2009). The ICT sector's own emissions are expected to increase, in a business as usual (BAU) scenario, from 0.53 billion tonnes (Gt) carbon dioxide equivalent (CO_2e) in 2002 to 1.43 Gt CO_2e in 2020. The ICT-enabled solutions would contribute to a savings of 1 ton per capita in 2020, if these materialize and are negated by the net growth of the total sector (Krishnan *et al.*, 2009).

In 2007, the total carbon footprint of the ICT sector—including (PCs) and peripherals, telecom networks and devices and data centers—was 830 Mt CO_2e, about 2% of the estimated total emissions from human activity released that year. Even if the efficient technology developments outlined in the rest of the chapter are implemented, this figure looks set to grow at 6% each year until 2020. The carbon generated from materials and manufacture is about one-quarter of the overall ICT footprint, the rest coming from its use (The Climate Group, 2008). So here a consumer awareness strategy might pay off way more than in other sectors where use is usually a smaller percentage of the total energy consumption.

The first decade of this new millennium has been marked, even taking the economic crisis into account, by an extraordinary economic productivity and an alarming decline in the viability of eco systems that are the foundation of our prosperity and survival. The simple truth is that we as a species are able to destroy our own conditions for prosperity. Industry and end-consumers use more resources than the earth can produce, and if we do not find alternatives we will be hard-pressed to find alternatives, if these are still around within 30–50 years, not to mention additional costs and loss of bio-productive resources that we incur because of pollution, climate change, and sea-level rises.

Electrical and electronic waste

Electrical and electronic waste (e-waste) is one of the fastest growing components of the global waste stream and, arguably, one of the most troublesome. The European Environmental Agency calculates that the volume of e-waste is now rising approximately three times faster than other forms of municipal waste. The total annual global volume of e-waste is soon expected to reach approximately 40 million metric tons—enough to fill a line of dump trucks stretching half way around the world. Rapid product innovations and replacement, especially in ICT and office equipment—the migration from analog-to-digital technologies and to flat-screen TVs and monitors, for example is fueling an increase of e-waste (United Nations University, 2007). For manufacturers, improving the e-waste recycling process is essential to continuity in business one or two decades from now. Unqualified or unscrupulous treatment of e-waste is still prevalent in many emerging economies and a lot of problems are exported by more developed nation states to developing countries. The inappropriate handling of e-waste leads, among others, to:

1. Emissions of highly toxic dioxins, furans, and polycyclic aromatic hydrocarbons.
2. Soil and water contamination from chemicals such as: brominated flame retardants (used in circuit boards and plastic computer cases, connectors, and cables); polychlorinated biphenyls (in transformers and capacitors); and lead, mercury, cadmium, zinc, chromium, and other heavy metals (in monitors and other devices). Studies show rapidly increasing concentrations of these heavy metals in humans; in sufficient dosages, these can cause neuro-developmental disorders and possibly cancer.
3. Waste of valuable resources that could be efficiently recovered for a new product life cycle. In many industrializing and developing countries, growing numbers of people earn a living from recycling and salvaging electronic waste. In most cases, though, this is done through the so-called backyard practices, often taking place under the most primitive circumstances, exposing workers to extensive health dangers (United Nations University, 2007). But what kind and what quantities of valuable resources are we talking about?

One metric ton of electronic waste from PC contains more gold than that recovered from 17 tons of gold ore. In 1998, the amount of gold recovered from electronic waste in the United States was equivalent to that recovered from more than 2 million metric tons of gold ore and waste. A ton of used mobile phones, for example, or approximately 6000 handsets (a tiny fraction of today's 1 billion annual production) contains about 3.5 kg of silver, 340 g of gold, 140 g of palladium, and 130 kg of copper, according to solving the e-waste problem (StEP). The average mobile phone battery contains another 3.5 g of copper. Combined value is over US$15,000 at today's prices (Electronic Takeback Coalition, 2010).

Another example: recovering 10 kg of aluminum via recycling, for example, uses no more than 10% of the energy required for primary production, preventing the creation of 13 kg of bauxite residue, 20 kg of CO_2, and 0.11 kg of sulfur dioxide emissions, and causes many other emissions and impacts. Compared to disposal, computer reuse creates 296 more jobs for every 10,000 tons of material disposed each year (Electronic Takeback Coalition, 2010).

In addition to well-known precious metals such as gold, palladium, and silver, unique and indispensable metals have become increasingly important in electronics. Among them, indium, a by-product of zinc mining, is used in more than 1 billion products per year, including flat-screen monitors and mobile phones. In the last five years, indium's price has increased six fold, making it more expensive than silver. Though known mine reserves are limited, indium recycling is so far taking place in only a few plants in Belgium, Japan, and the United States. Japan recovers roughly half its indium needs through recycling.

The market value of other important minor metals used in electronics such as bismuth (used in lead-free solders) has doubled since 2005, while ruthenium (used in resistors and hard disk drives) has increased by a factor of seven since early 2006. The large price spikes for all these special elements that rely on production of metals like zinc, copper, lead, or platinum underline that supply security at affordable prices cannot be guaranteed indefinitely unless efficient recycling loops are established to recover them from old products (United Nations University, 2007). Valuable resources in every wasted product with a battery or plug—computers, TVs, radios, wired and wireless phones, tablets, e-readers, MP3 players, navigation systems, microwave ovens, coffee makers, toasters, hair-dryers, to name but a few—are being trashed in rising volumes worldwide. Worse, items charitably sent to developing countries for re-use often ultimately remain unused for a host of reasons, or are shipped by unscrupulous recyclers for illegal disposal. And, e-waste in developing countries is incinerated, not only wasting needed resources but adding toxic chemicals to the environment, both local and global (United Nations University, 2007). What is currently done to curb the negative side effects of waste disposal of electronic equipment.

The European Directive on the restriction of the use of certain hazardous substances in electrical and electronic equipment (2002/95/EC) aims to curb the use of toxic substances in ICT equipments. The Guidelines on Waste Electrical and Electronic Equipment (2002/96/EC), WEEE in short, aims specifically at reducing ICT waste (Lieshout and Huygen, 2010). The WEEE is interesting because it makes ICT producers to a certain extent responsible for recycling their products, i.e., it aims at internalizing the environmental costs of the ICT industry (see also Chapter 5).

An industry-led efficiency approach promises to have quite some impact and could help us move globally to a low carbon economy. The Climate Group report (2008) mentions that specific ICT opportunities can lead to emission reductions five times the size of the ICT sector's own footprint, up to 7.8 Gt CO_2e, or 15% of total BAU emissions by 2020.

Reuse happens but not systematically and given the short innovation cycles in the industry, hybridization of products is a trend that is emerging. Partial reuse, another word for hybridization, caters to consumer preferences because exteriors can be replaced while the technology driving the device is long-lasting and will be taken back by the producer to retain and reuse the resources embedded in it. The trick is to make the different components easily separable, save costs, and create conditions to upgrade the resources used. Currently most of what is reused is donated and in the end thrown away in developing countries that do not tend to have good e-waste recovery infrastructures. All the more reason to design products that can be recycled or reused with as little energy as possible or, alternatively, can easily be separated into components that can be reused or recycled. Such designs would likely also increase labor participation in developing countries under much healthier circumstances and opens up the possibility to automate disassembly and reuse/recycling, spurring a new round of wealth creation.

It is clear that the global ICT industry has chosen an efficiency strategy, with some eco-effective elements. Standardizing recycling processes globally to harvest valuable components in electrical and electronic scrap (e-scrap), extending the life of products and markets for their reuse, and harmonizing world legislative and policy approaches to e-scrap are the prime goals of a new global public–private initiative called StEP. Major high-tech manufacturers, including Hewlett-Packard, Microsoft, Dell, Ericsson, Philips, and Cisco Systems, join UN, governmental, NGO, and academic institutions, along with recycling/refurbishing companies as charter members of the initiative (United Nations University, 2007) in which reduction in the form of dematerialization, recycling, and reuse, to recoup precious resources, and regulation to induce compliance with standards are the main foci of attention. Especially attention to recycling and reuse combined with standardization can produce quick results and put the whole industry on a path toward eco-effectiveness. To beat Jevons Paradox, more is needed but we have to start and start learning how we can improve the bottom-line for companies, environment, and employees. Because standardization seems one of the most promising policies, we devote a section to this topic mainly to sketch the state of the art.

1.3   STANDARDS

In Chapter 5, standardization features prominently, and here we draw some attention to a number of standards that are already around. Then we further summarize and elaborate the concept of eco-effectiveness.[2]

Environmental (business) standards
Several business standards have been developed to assess various aspects of an organization's operations. Total Cost of Ownership (TCO) and Life Cycle Costing (LCC) are methods used in IT investment valuation (Ellram, 1995). Life Cycle Analysis (LCA) is also a widespread, standardized method often used by organizations (ISO, 2006; Zuthi and Sohal, 2003). Next to sustainability assessment tools at product level, several business standards have been developed to assess the impact of an organization's activities. These standards often address one or more environmental issues. We discuss three business standards. First, the greenhouse gas (GHG) protocol will be discussed because it is the most widely used emission standard (Kolka *et al.*, 2008). Second, the water footprint introduced by Hoekstra and Hung (2002) and further elaborated by Chapagain and Hoekstra (2003) will be outlined as it provides useful insight into organizational water use. Third, ways of accounting for raw materials, energy, and solid waste will be discussed briefly as incorporated in the ISO 14051 material flow cost accounting. These standards have been used to understand how the framework can be designed in terms of setting organizational measurement boundaries (i.e., the relative greenness of an organization's hardware IT infrastructure) and performance indicators.

GHG protocol
The WRI and WBCSD have developed the GHG protocol. The GHG protocol has become a global accounting tool used by numerous governments and businesses around the world to understand, quantify, and manage GHG emissions. The protocol is the most widely used standard (Kolka *et al.*, 2008). In the GHG protocol, a distinction is made between direct and indirect GHG emission. Direct emission is caused by sources that are owned or controlled by the organization and the indirect emissions are a result of activities of the companies that occur at sources owned or controlled by another company. The scope of an organization's GHG emission has been divided into three classes: direct GHG emissions, electricity indirect GHG emissions, and other indirect GHG emission. The direct GHG emission (scope 1) occurs from sources owned or controlled by the organization, such as furnaces and vehicles. Electricity indirect GHG emission (scope 2), accounts for GHG emission from the generation of purchased electricity used by the organization. This also included transmission and distribution losses. Other indirect GHG emissions (scope 3) occur from sources not owned or controlled by the company. Examples of such are extraction and production of purchased materials and fuels, leased assets, franchises, and outsourced activities and waste disposal (WBCSD and WRI, 2004). Furthermore, in March 2012 the WRI and WBCSD published the draft version of the "GHG Protocol Product Life Cycle Accounting and Reporting Standard—ICT sector guidance." The protocol builds on the International Standard Organization (ISO) LCA standards 14040:2006 and 14044:2006 and provides a consistent and pragmatic approach to GHG assessment of ICT products (GHG Protocol, 2012; WBCSD and WRI, 2012).

Water footprint
Besides the business standard for determining GHG emissions of organizations, a water footprint standard has been developed as an analogy to the ecological footprint. The water footprint is linked to the virtual water concept which is the volume of water necessary to produce a commodity or service (Hoekstra and Chapagain, 2007). The water footprint of nations developed by

---

[2]This section has largely been taken from Chapter 7 of this book. We would like to thank the authors Kalsheim & Beulen for their kind permission to use this section at a more appropriate place.

Hoekstra and Chapagain (2007) makes a distinction between volume of water used from domestic water resources and water used in other countries to produce imported and consumed goods and service. The water footprint of an organization is the total volume of fresh water consumed or polluted directly or indirectly to run and support the business which is expressed in water volume per unit of time. The business water footprint consists of two components: the operational water footprint and the supply chain water footprint. Hoekstra *et al.* (2009) additionally distinguish between three types of water: green, blue, and gray. The concept of green water entails rainwater evaporated from the soil during the production process of agricultural products. Blue water is referred to as the volume of freshwater that evaporated from the global blue water resources (surface water and ground water) to produce the goods and services consumed by the individual or community. Gray water is defined as the amount of polluted water as a result of business activities (Hoekstra *et al.*, 2009).

Raw material cost accounting
Raw materials use in business accounting often focuses on product chains only, whereas a company's supply chain also includes providers of products, processes, technology, and services that are not immediate raw materials of a company's core products (Drury, 2007; Kovács, 2008). Nevertheless, in organizational accounting standards such as the recently introduced ISO 14051 standard, the material flow accounting approach has been implemented which incorporates raw materials and soiled waste streams. The flows and stocks of materials within organization are traced and quantified, and the costs associated with those material flows are incorporated in the evaluation of material flows. This includes raw materials as an input and solid waste as an output (Jasch, 2000). In a broader view, materials flow is defined as the course of a material from its extraction to its disposal. As part of organizational decision making, this is often narrowed down to corporate level, incorporating supplier input and customer and disposer outputs. In materials flow cost accounting, economic and ecological objectives are integrated in order to contribute to a reduced or more efficient use of materials. Costs that have been incorporated in this standard are related to materials costs, system costs such as maintenance costs and waste management costs.

1.4   DISCUSSION: EFFICIENCY AND EFFECTIVENESS PRESUPPOSE EACH OTHER

There are a lot of signals that BAU is over. BAU ultimately destroys the natural base on which we built all our wealth. We showed that awareness has translated in global action; the ICT industry puts its sustainability bets on an eco-efficiency strategy. The fact remains however that the framing of the problem leads to solutions that follow the credo: "do more with less resources and energy." Throughout the book you will see that this is still the predominant generic strategy to become more sustainable. We strongly support the move the industry worldwide is making in the area of (energy) efficiency, because it gives relief to a host of problems we will face in the coming years. But an efficiency strategy provides just that, relief. Population and economic growth easily cancel out efficiency gains of 15–20% and that is what is called the Jevons Paradox or rebound effect. Canceling out the rebound effect necessitates a behavioral change at the user side and a change in production regimes at the industry side, while it challenges knowledge institutes to come up with way more innovative solutions than are currently offered and it challenges governments and regulators to come with institutional frameworks that reward environmentally conscious behavior in business arenas, labor and consumer markets.

Solutions that need to be found will require more than becoming even more efficient. Systemic change also requires change at the institutional level, usually the domain of politics in a co-production with regulators. However, even if the political field would do nothing, industry will need to alter the ways in which they design and produce in order to remain flourishing as a sector, exactly because some raw materials are increasingly getting scarce.

A number of ecologically intelligent design approaches to architecture, industry, and architectures that involve materials, buildings, and patterns of settlement have been formulated and developed over the past two decades. Elements of CtoC are also found in industrial ecology (Ehrenfeld, 1997) and biomimicry (Benyus, 2002) and both approaches were formulated from an engineering perspective. In the agricultural field deep ecology, as a philosophy, a new radical way of redesigning agriculture, permaculture, spawned in such a way that it functions once again as a CO_2 sink, contributing to instead of destroying (bio)diversity and is more productive per square meter.

These approaches are labelled as eco-effective because these try to support a transition to an economy and a society that is not harmful, excluding and economically unfeasible in the long run. The keyword is the generation of value: eco-effective systems tend to generate value in more than one domain and once installed for a long period of time. The main thing is that you go further than reduction of energy and resource use, the rationale underlying all efficiency approaches.

Given the challenge to cancel out the rebound effect, we conclude that eco-efficiency by itself is, as a strategy, necessary but insufficient. Eco-effective strategies should also be formulated and synergies will develop because the strategies are mutually supportive. These almost presuppose each other, because eco-efficiency strategies generate funds in the form of cost savings that should be reinvested in eco-effective design of infrastructures and appliances to ensure a sustainable future for the sector. Efficiency further supports the move to an effectiveness strategy because the more energy efficient a product is, the easier it becomes to design and produce eco-effective products. Simply because alternative sources of energy generation and supply become feasible and the software and technical support, that help ensure a stable supply of energy are increasingly available. We also put forward that following up: an eco-efficient strategy with an eco-effective strategy is good business sense because it creates new markets as can currently be witnessed in the energy sector.

REFERENCES

Benyus, J.M.: *Biomimicry: innovation inspired by nature.* William Morrow Paper-backs, 2002.

Chapagain, A. & Hoekstra, A.: Virtual water trade: a quantification of virtual water flows between nations in relation to international trade of livestock and livestock products. In: *Virtual water trade, Proceedings of the International Expert Meeting on Virtual Water Trade,* 2003.

Electronic Takeback Coalition: Facts and figures on e-waste and recycling, 2010, http://www.electronics takeback.com/wp-content/uploads/Facts_and_Figures (accessed February 2010).

Drury, C.: *Management and cost accounting.* Cengage Learning EMEA, 2007.

Ehrenfeld, J.R.: Industrial ecology: a framework for product and process design. *Journal of Cleaner Production* 5:1 (1997), pp. 87–95.

Ellram, L.M.: Total cost of ownership: an analysis approach for purchasing. *International Journal of Physical Distribution & Logistics* 25:8 (1995), pp. 4–23.

GHG Protocol: Calculation tools. 2012, http://www.ghgprotocol.org/calculation-tools/all-tools/ (accessed July 2012).

Hoekstra, A. & Chapagain, A.: Water footprint of nations: water use by people as a function of their consumption pattern. *Water Resource Management* 21:1 (2007), pp. 35–48.

Hoekstra, A., Chapagain, A., Aldaya, B. & Mekonnen, M.: The water footprint manual. Water Footprint Network, Enschede, The Netherlands, 2009.

Hoekstra, A. & Hung, P.: Virtual water trade: a quantification of virtual water flows between nations in relation to international crop trade. *Value of Water Research Report Series* no. 11, IHE. Delft. 2002.

Huijbregts, M.A., Hellweg, S., Frischknecht, R., Hendriks, H.W., Hungerbuhler, K. & Hendriks, A.J.: Cumulative energy demand as predictor for the environmental burden of commodity production. *Environmental Science & Technology* 44:6 (2010), pp. 2189–2196.

Huijbregts, M.A., Hellweg, S., Frischknecht, R., Hungerbühler, K. & Hendriks, A.J.: Ecological footprint accounting in the life cycle assessment of products. *Ecological Economics* 64:4 (2008), pp. 798–807.

ISO Standard. 14040: Environmental management—life cycle assessment—principles and framework. British Standards Institutionm London, UK, 2006.

Jasch, C.: Environmental performance evaluation and indicators. *Journal of Cleaner Production* 8:1 (2000), pp. 79–88.

Kokubu, K., Kos, M., Furukawa, Y. & Tachikawa, H.: Material flow cost accounting with ISO 14051. *ISO Management System* 9:1 (2009), pp. 15–18.

Kolka, A., Levyb, D. & Pinksea, J.: Corporate responses in an emerging climate regime: the institutionalization and commensuration of carbon disclosure. *European Accounting Review* 17:4 (2008), pp. 719–745.

Kovács, G.: Corporate environmental responsibility in the supply chain. *Journal of Cleaner Production* 16:15 (2008), pp. 1571–1578.

Krishnan, S., Balasubramanian, N., Subrahmanian, E., Arun Kumar, V., Ramakrishna, G., Murali Ramakrishnan, A. & Krishnamurthy, A.: Machine level energy efficiency analysis in discrete manufacturing for a sustainable energy infrastructure. *Proceedings of the Second International Conference on Infrastructure Systems and Services: Developing 21st Century Infrastructure Networks (INFRA 2009)*, IEEE, 2009, pp. 1–6.

Lieshout, M. V. & Huygen, A.: ICT en het milieu Mag het een bitje meer? In: V. Frissen & M. Slot (eds): *Jaarboek ICT en Samenleving 2010, 7de editie: De duurzame informatiesamenleving*. Media Update, Gorredijk, The Netherlands, 2010, pp. 139–158.

Mintzberg, H.: *Managing*. Prentice Hall/Financial Times, 2009.

NASSCOM: Green Warriors Newsletter. 2009, http://www.greenwarriorinc.com/ (accessed April 2013).

Smart 2020: Enabling the low carbon economy in the information age. 2008, The Climate Group, http://www.smart2020.org/publications/ (accessed, march 2010).

United Nations University: UN, industry, others partner to create world standards for e-scrap, 2007.

WBCSD & WRI: GHG protocol: a corporate accounting and reporting standard. World Resources Institute and World Business Council for Sustainable Development, Washington, DC, 2004.

WBCSD & WRI: GHG protocol product life cycle accounting and reporting standard—ICT sector guidance. World Resources Institute and World Business Council for Sustainable Development, Washington, DC, 2012.

Zuthi, A. & Sohal, A.S.: Adoption and maintenance of environmental management systems: critical success factors. *Management of Environmental Quality: An International Journal* 15:4 (2003), pp. 399–419.

CHAPTER 2

Creating synergies between approaches and tools for sustainable ICT development

Jaco Appelman, Freek Bomhof & Reinier de Nooij

2.1 INTRODUCTION

As more people start to recognize the need for a transition to more sustainable systems, there has been a proliferation of approaches, methods, and tools to support such change (Edwards, 2005). However, the myriad of approaches and tools is known to be slowing the process toward sustainability rather than assisting it. "Discussions about which framework is right and energy spent on battling each other's positions can lead to deadlocks" (Senge *et al.*, 2007). This is quite a deep observation by Senge because it brings the focus from the object to be changed, the unsustainable world, to the people and organizations that try to change it. Despite all good intentions, they hinder roads to sustainability rather than assisting it. Instead of arguing which approach is right or the best, we are looking for combinations, preferably with synergistic properties to support a transition we are trying to make.

Synergy means, in this chapter, that the combination of approaches works better than applying them independently or consecutively. This may be because an omission in one approach is filled by another which means that a disadvantage of one approach is solved by another or there is the possibility that two qualities of different approaches reinforce each other. The rationale to pursue a combination of approaches is that there is an increasingly pressing need to support a transition to sustainability.

The goal is therefore to see if a combination of existing, science-based, increasingly well-known sustainability approaches can, theoretically, be sufficient to cover all requirements for a sustainability transition in a regional, cross-sector, multi-stakeholder context and, based on this overview, where and how synergies could be created. To reach our goal, we aim to answer the following questions in this chapter:

1. What existing influential or well-known approaches are suitable within in a local and regional cross-sector multi-stakeholder context?
2. How do these approaches relate to the requirements for processes leading to a sustainability transition?
3. Is a combination of these approaches sufficient to cover these requirements?

In this way we hope to lend more substance to the goal of transition management, a particular, institutional branch of change management (Rotmans *et al.*, 2001). To compare the different sustainability approaches, we need a set of selection criteria and a common frame of reference. This will be the subject of Section 2.3. In Section 2.2, we provide a background to the approaches selected. In Section 2.4, the frame is built from literature, an expert meeting and our own experiences. This analysis is not exhaustive and open to new input, but it should suffice to give guidance to complex change trajectories. The chapter ends with general conclusions and specific conclusions for the ICT sector.

2.2   SELECTION OF APPROACHES

We selected approaches or concepts that:

- have publications in scientific literature,
- are not limited to a specific discipline (e.g. architecture) or environmental problem (e.g. climate change), and
- attempt to come up with a set of principles that could serve as a guide or reference points when one wants to embark on a transition towards sustainability.

Projects, manifestos, principles, institutes, and curricula are excluded. Therefore, although extremely interesting and certainly part of a larger movement to make a transition toward a more balanced planet: resilience-movement, transition towns, permaculture inspired movement, zeronauts, 1% club, etc. are excluded from this analysis.

Based on these criteria we initially selected the following approaches:

- The natural step (TNS)
- Cradle to cradle (CtoC)
- Ecological footprint (EF)
- Biomimicry (B)
- Industrial ecology (IE)
- Natural capitalism (NC)
- Corporate social responsibility (CSR)

In further discussions we wanted to present recognizable or relatively well-known approaches among sustainability professionals, policymakers and the general public in The Netherlands and seems intuitively complementary:

- The natural step (TNS)
- Cradle to cradle (CtoC)
- Ecological footprint (EF)

Why we contend that the three approaches might complement each other is that all three take the earth geologically as a refererence point irrespective of business or any other form of social organization. Taken together and because of this irrespectiveness, we also feel that they are eminently suitable to provide guidance at multiple levels in private and public settings.

We acknowledge that industrial ecology (Ehrenfeld, 1997) and biomimicry (Benyus, 2002) are design approaches that depart from a similar normative framework in the sense that we as humans (part of nature) should learn from nature how to design our products and production systems. Both biomimicry and industrial ecology are just as CtoC essentially design and analytical approaches and we hope the reader accepts that CtoC functions as an example of that stream of sustainability approaches.

CSR predominantly aims to support business, not public organizations or public–private partnerships, and our focus was on societal systems, that include economic actors. But we are open to the possibility that CSR does inspire and inform certain aspects of sustainability that might be missed by our collection.

Natural capitalism proposes a new approach to economics and its political economic agenda is therefore very clear but aimed at the institutional level. Although we need a regulatory environment that supports a transition toward sustainability, we do not aim to inform, solely, the political arena but people that need to either decide about or implement actual programs or projects, national or regional policies; we therefore disregarded natural capitalism. To support the reader into coming to grips with the subject matter we introduce, as a background, the three main approaches studied that will be analyzed.

2.2.1 Description of approaches selected

The natural step

TNS is a systemic approach for sustainability, developed in Sweden by Karl-Henrik Robèrt and his colleagues, and applied in businesses and local communities in more than a dozen countries. The objective of TNS is to create more solid platforms for strategic decision making through systems thinking and dialogue. TNS describes unsustainability as a funnel metaphor, formed by the systematically increasing demands on resources and the systematically declining capacity of earths ecosystems to meet these demands. The system conditions describe what is required to balance demand and supply. With those basic principles organizations can, theoretically, guide themselves to:

- analyze current practices from a sustainability perspective,
- create visions and solutions within the same perspective, and
- elaborate strategic step-by-step action programs (backcasting).

In their current form the principles or system conditions (SCs) of TNS are the following:

In a sustainable society, nature is not subject to systematically increasing. . .

1. concentrations of substances extracted from the earths crust (such as heavy metals and fossil CO_2 in the atmosphere),
2. concentrations of substances produced by society (such as POPs and NO_x),
3. degradation by physical means (such as deforestation and overfishing), and
4. people are not subject to conditions that systematically undermine their capacity to meet their needs (such as unfair trade and others ways of abuse of economic or political power) (Robèrt, 2000; Robèrt *et al.*, 2002).

Cradle to cradle

CtoC (McDonough and Braungart 2002a) is an ecologically intelligent approach that involves materials, buildings, and patterns of settlement which are wholly healthful and restorative. It is based on the work of Stahel (1982), whose inspiration can be traced back to Chapman (1974), and Hirst (1974). CtoC design sees human systems as nutrient cycles in which every material can support life. Materials designed as biological nutrients provide nourishment for nature after use; technical nutrients circulate through industrial systems in closed-loop cycles of production, recovery and remanufacture. Following a nascent science-based protocol for selecting safe, healthful ingredients, cradle-to-cradle design maximizes the utility of material assets. Key concepts are:

1. Use only renewable energy sources such as solar power, hydropower, and wind power.
2. There is no waste, only food. Develop products and processes that mimic natural systems. Outputs of any component in the system should provide nutrients for the biosphere or the technosphere, and no loss of quality should occur.
3. Respect and cherish diversity. This includes diversity of life, culture, place, needs, and uniqueness of people.
4. Promote the rights of people and nature to coexist in a healthy, mutually supporting diverse and sustainable situation.

Ecological footprint

The first academic publication about the EF was by William Rees in 1992. The EF concept and calculation approach was developed by Wackernagel and Rees (1996). EF analysis compares human demand on nature with the capacity of the biosphere to regenerate resources and provide services. It does this by assessing the biologically productive land and marine areas required to produce the resources a population consumes and absorbs the corresponding waste, using current technologies. This resource accounting is similar to life cycle analysis wherein the consumption of energy, biomass (food, fiber), building material, water, and other resources are converted into a normalized measure of land area called "global hectares". Per capita EF is a means of comparing consumption and lifestyles, and checking this against nature's ability to provide for this consumption.

The approach can inform policy by examining to what extent a nation uses more (or less) than is available within its territory, or to what extent the nation's lifestyle would be replicable worldwide. The footprint can also be a useful tool to educate people about carrying capacity and overconsumption, with the aim of altering personal behavior. EFs show that many current lifestyles are not sustainable. Such a global comparison also clearly shows the inequalities of resource use on this planet at the beginning of the twenty-first century. As such, although it is essentially an accounting approach, the truth it reveals about how we deal with our planet and fellow human beings immediately gives rise to ethical and political considerations. The focus of EF is therefore primarily aimed at public organizations and governmental bodies at all levels and influencing the institutional rules. Although this is changing and business is increasingly using carbon and water footprints to legitimize and improve their operations.

After this short introduction, we turn to the question: how to compare them? What requirements would need to be fulfilled to ensure a transition toward more sustainable production and living?

2.3   CREATION OF FRAME OF REFERENCE

In this section, requirements for sustainability processes to be met are gathered from literature. Selection of the authors is neither exhaustive nor completely random. We selected authors that have a longstanding reputation in their respective fields and have discussed or researched sustainability as a subject matter.

2.3.1   *Requirements for sustainability transitions*

The still most quoted definition of sustainable development is the one formulated in *Our common Future*, the UN-report made the concept a meme. *"Sustainable development is development which meets the needs of the present without compromising the ability of future generations to meet their own needs"* (United Nations, n.d.). This definition provides no criteria that allow for a comparison of sustainability approaches. However, it does make clear that sustainable development is about supporting and increasing quality of life for all species over time. The needs of different species, groups, and different generations needs to be balanced. Sustainable development is therefore an inherently normative and political concept.

A transition is a set of connected changes, that reinforce each other but take place in several different areas at different moments in time, such as technological, economic, institutional, behavioral, cultural, ecological, and belief systems (Rotmans *et al.*, 2001). The enormous complexity often frustrates the execution of sustainability programs. Research shows that the more complex the situation, the more important psychological and cultural aspects are (NLRO, 1999). This implies that the process design of a sustainability approach must explicitly address these aspects.

Transition management is an example of a change management approach that aims to alter institutional rules; that pushes for a transition. The approach describes the structure of organizing a process of change; of which the basic steps or elements are:

- Organizing a multi-actor network;
- Developing sustainability visions and transition agendas;
- Mobilizing actors and executing projects and experiments; and
- Evaluating, monitoring, and learning.

More detailed process designs need to be contextualized because sustainability transitions require applicability of approaches at different levels: local and regional, multiple stakeholder, multi-sector, and multilevel. Furthermore, a transition process has its own temporal dynamics. It may be in a start-up, acceleration, or declining phase, each phase requiring a different approach (Rotmans *et al.*, 2001). In each phase there are four categories of key contextual issues that need to be addressed in change programs for sustainability:

1. Individual subjective factors (values, worldview, emotions, perceptions, etc.)
2. Individual objective factors (socio-demographics, skills, knowledge, role, etc.)
3. Collective subjective factors (culture, shared norms, regimes of denial, and acknowledgment)
4. Collective objective factors (political, economic, technological rules, regulations, and laws) (Ballard, 2005)

Each factor may either hinder or facilitate a sustainability transition. It is therefore advisable to consider all these factors when planning for sustainability. But we should be aware to what extent it is possible to influence all factors. We mentioned that we excluded the institutional level because the accent lies on engendering change at the local and regional level, not on the steps needed to create an institutional environment conducive to sustainable change. We will therefore not pay attention to the collective objective factors. But for the fact that it goes without saying that the current collective objective factors, to a large extent, perpetuate our harmful ways of production and consumption (Jackson, 2009). In order to address the remaining three factors, we follow Ballard who emphasizes that change always needs to entail:

1. "awareness" of what is happening and what is required;
2. "association" with other people in groups and networks to reach the requirements; and
3. "agency" or the ability to find a response that seems personally meaningful (Ballard, 2005).

The conclusions of Ballard are echoed by Senge *et al.* (2007), who describes the types of work required to attain these three conditions; successful collaborative efforts must embrace three interconnected types of work: conceptual, relational, and action driven. Conceptual work is related to awareness, association with relational work, and action driven work is easily linked to the concept of agency.

Meeting these three conditions requires a process, a process of action and reflection. This means actions are reflected upon and changed when needed: Action and reflection is also necessary in integrating work across the three conditions. In order to attain the three conditions for change, that is awareness, association, and agency: collaboration is key.

Senge *et al.* (2007) emphasizes that successful collaborative efforts must embrace these three interconnected types of work: conceptual, relational, and action driven. We elaborate on them to further uncover what requirements are needed to come to an effective change management approach. *Conceptual* work is about framing complex issues in a way that makes them understandable, creating clarity without denying complexity, simplicity without reduction. This requires a systems perspective (Senge *et al.*, 2007). Systemic thinking tends to promote radically innovative action. The systems perspective broadens the scope for many actors and enables sustainability practitioners to see the various interconnections. It also enables them to take a broad perspective necessary to transcend organization bound goals. This helps keeping a large number of options open. We tentatively conclude that conceptual work is important for creating awareness, but also for association and agency. *Relational* work creates a foundation for cooperation, trust, mutuality, and joint learning. It requires approaches that make people inspired and feel like they want to be part of this process. Examples of approaches are reflective conversation, working with mental models and the World caf: collaborative dialogue and knowledge sharing. Here it is important to get the system in the room. People thinking and planning together will add to each others awareness of the subject and get a clearer sense of what meaningful actions are. In other words, relational work aims at creating association and, if successful, adds to the agency and awareness of the group and its members.

Action driven work means that concrete plans are made with the intention to carry them out. This will bring focus through a summation and visualization of a list of meaningful actions and

Table 2.1.   Comparison of approaches and concepts.

Requirement	TNS	CtoC	EF	Sufficient?
Issues addressed in scope				
Social	2	2	1	Y
Ecological	2	2	2	Y
Economical	2	2	0	Y
Political/institutional	1	0	2	Y
Ethical	2	2	2	Y
Psychological/spiritual	1	0	0	N
Physico-chemical	2	2	2	Y
Biological	2	2	2	Y
Socio/cultural	2	1	0	Y
Economical	1	1	0*	Y
Measurement tool	1	0	2	Y
Contextual factors addressed				
Psychological	1	1	0	N
Behavioral	2	2	2	Y
Cultural	1	1	0	N
Systems	2	2	2	Y
Process: conditions and types of work				
Awareness (conceptual)	2	1	2	Y
Association (relational)	1	1	0	N
Agency (action driven)	2	2	2	Y
Action and reflection process	2	0	0	Y
Shared understanding	2	1	1	Y
Vision	2	2	1	Y
Inspiration	1	2	1	Y
Passion	1	2	0	Y
Leadership	2	1	0	Y
Positive approach	1	2	0	Y

a time frame. When you do this with a group people, they tend to become more aware how the problem is structured and what the limits of human agency are. It can also lead to new forms of association, necessary for getting down to action, to be an agent of change.

Now where does this lead us to? It gives a first indication that we need to do a lot at every level of a social system and the question arises if such an approach is feasible in its entirety. If we can drive holistic change management programs or need to come up with a more fine grained sub-subdivision. We therefore proposed in this section a more easy to understand and use subdivision at the learning and process level. We found that a process design must show how awareness by means of conceptual work, association by relational work, and agency by action driven work is created, at the individual and group level. When these conditions are met things will change. Change for the sake of change is not enough and we also need to ensure that the change is aimed at creating a more sustainable system, the normative side of the coin. Before we go into values, we relate the outcomes of an expert-session on exactly the same topic as we just discussed from the literature. The set of criteria derived from literature (Table 2.1) is checked and supplemented with the outcomes of the expert meeting. In this way we have performed an expert validation that gives us information of the face validity of the combination of approaches and helps us identify possible gaps.

2.3.2   *Expert meeting*

In the spring of 2009 a meeting was organized with sustainability experts that were involved in either the TNS, CtoC, or EF[1]. Each participant worked mainly with one of these three approaches. The goal of the meeting was to create consensus on the requirements for sustainability transitions and exploring possibilities for synergies at the process level between the three approaches. In this section, we will explain the design of the meeting briefly and present the outcomes that were instrumental in the creation of the framework

Design of the expert meeting

In order to capture as much information as possible during one day, several technologies were used: a group support system (computerized group decision room), whiteboards, flip-overs with post-its and soft walls. With the group support system, we could ensure that the three groups did not influence each other initially, the information given could be easily separated along the lines of the three participating groups (i.e. TNS, EF, and CtoC) and we could display a group memory when needed; to help the group to remain focused. We combined insights from group model building with collaboration engineering to come to a design that would deliver a lot of information in a short time.

All participants were divided into pairs consisting of two members of the same group[2] and given a laptop computer. Supported by a facilitator and a computer, that displayed the questions that were introduced by the facilitator, answers were entered into the group decision room by each pair simultaneously. Each pair was given the same amount of time to answer each question. When the allotted time had passed, the next question appeared on screen.

During this part of the session, four questions were asked:

1. What is the most important issue in sustainability transitions?
2. What are the most important barriers in transitions?
3. What is required to overcome these barriers?
4. What will create the movement necessary for the sustainability transition (initiation)?

All the answers were printed on small papers. These prints were then clustered by the group on metaplan boards in the second part of the session. In the third part of the session, each approach was presented to the rest of the group in order to get everyone acquainted with all three approaches. In the discussion afterward, the various possibilities for combining the approaches were explored.

Results expert meeting

Table 2.1 shows that the important *issues* according to the experts are inspiring vision, coop-eration (based on community feeling and purpose), awareness, clear science-based criteria for sustainability, and systemic change[3]. CtoC experts gravitated towards inspiring visions, the TNS group emphasized cooperation and the experts working with EF focused on clear science-based criteria for sustainability.

The most important *barriers* in sustainability transitions as perceived by our group of experts are lack of consensus on a systematic science-based learning process design (emphasized by the TNS group), psychological and cultural factors (emphasized by CtoC experts), and the way our political and economic systems are organized (emphasized by EF).

The opinions on *what is required* is to overcome these barriers stress the importance of awareness, communication, cooperation, changing the financial and tax system, and innovation, leadership, and passion. The CtoC group values innovation, leadership, and passion most of all. EF experts focus on the financial and tax system, whereas the TNS group centres on awareness.

---

[1]The group involved 28 experts, nearly all senior members of governmental institutes, colleges, universities, NGOs, and companies dedicated to sustainable development.
[2]There were five EF pairs, three CtoC pairs, and six TNS pairs.
[3]The issues that have been italized in the table are the issues brought forward by the group.

With regard to what is *necessary to create/initiate movement* a sustainability transition, the group of experts mentions the importance of a positive approach, involvement in creation, inspiration, passion, and action, and the importance of finding and using synergy between the three approaches. The experts working with EF focused on adopting a positive approach that rewards low footprints whereas CtoC expert believe in inspiration, passion, and action. The TNS group emphasizes involvement in cocreation.

Discussion
The results of the expert meeting provide an extra underpinning for the set of criteria derived from literature.

The findings show that the group of experts also distinguished between the categories "awareness", "association", and "agency". The categories of barriers known from the literature were also produced by the group with one extra; the group put psychological and cultural barriers in a separate cluster (Tables 2.2–2.5, see Appendix). A need for clear criteria and, preferably unambiguous, metrics and measurement tools was mentioned much like a GDP now functions to measure economic activity, an aspect lacking in the change-management and transition literature where metrics and measurement are deemed to be a negotiated reality. Additional requirements for sustainability transitions resulting from the expert meeting are:

- Inspiring vision
- Leadership, passion
- Positive approach: tempting & rewarding instead of doom and control.

It can be argued that the three requirements above result from combinations of "awareness", "association", and "agency", e.g. awareness and agency gives inspiration, all three combined presuppose the existence of leadership and passion. In the following analysis, however, these requirements are initially treated as autonomous and added to the analytical frame.

Consensus on possibilities for synergistic combinations
The group of experts agreed that it is possible and desirable and valuable to combine the three approaches. It was concluded that:

- TNS could serve as a strategic framework and process required to implement the vision presented by CtoC and that EF is a tool for measuring baseline conditions and progress.
- The enthusiasm generated by CtoC, or other design approaches, is invaluable, maybe indispensable, as a complement to the cognitive awareness created by TNS and EF.
- The emphasis on eco efficiency by EF is a necessary complement to eco-effectiveness as promoted by CtoC. But it only works if the institutional environment favors more sustainable solutions.

To connect the expert meeting and literature study, we compare the outcomes in more detail in the following section. We do this by first presenting the results in a condensed form, then we return to the three approaches and provide short case studies to highlight the strengths and weaknesses of each approach.

2.4 ANALYSIS AND COMPARISON OF TNS, CtoC, AND EF

2.4.1 *Analysis overview*

The three approaches are compared with respect to the criteria distilled from the literature and the expert meeting. This is based on the literature and a case study for each approach. The approaches are scored based on how well they match the criteria. Three scores are distinguished:

- 0: approach does not match the criterion. This means an issue is lacking or the approach is not suitable in one of the contexts specified.

- 1: approach matches the criterion but not sufficiently. This is the case if an issue is only implicitly addressed, or not in a science-based way.
- 2: approach matches the criterion. This is the case if an issue is explicitly addressed, in a science based way.

Science-based means that the approach studied specifies how we can measure (unambiguously?) the attainment of the goals of sustainable development. We allow for the fact that much of our "objective" knowledge is socially mediated or negotiated, at the same time things like energy used, CO_2 produced, or resources involved in the production of an item like a PC can be expressed as stable enough figures that do not change much over time.

We revisit our descriptions of the three approaches and elaborate on them while also providing more insight by discussing a case study for each approach. In the conclusions, we then concern ourselves what combinations of approaches might be synergistic.

But we begin with the presentation of the combined table. In the following sections, we elaborate on the condensed findings represented in the table. We start with what they have in common: each approach starts with the recognition that we need to find ways to alter our behavior and thinking in order to maintain a decent living for everyone, preferably elegantly enjoyed. Each of the approaches emphasizes the urgency of the matter but none has provided clear answers how to motivate people. The approaches themselves were or course born out of a desire to motivate people to start doing things in different ways but none of the approaches explicitly addresses the psychological level and or behavioral level. They also have their own institutes and a lot of spin-offs in terms of companies that operate in an accredited or unaccredited fashion. Leadership refers to leadership within an approach, and each of these has its own recognized leader or set of leaders who also tend to be the first visionaries. At the same time any process of change at group or higher levels will require other leaders and leadership. This is what the participants emphasized. We will relate and connect our main findings through a discussion of each approach and a short case from each of the three approaches.

2.4.2   *The natural step*

TNS has an explicit process design and aims to provide a common framework of reference for a diverse group of stakeholders, the approach is highly generic and therefore applicable in a multisector and multilevel setting. Economic aspects are addressed in the problem analysis and the process design (funnel, D-step, return on investment, guidebook: business case), but no tools are offered. The system conditions only provide a basis for a measurement tool (guide). The framework does not offer a measurement tool but suggests using life cycle analysis (LCA) and material flow analysis (MFA) and/or EF (personal commun.: K.H. Robert, 2011).

TNS has been developed based on a thorough understanding of "system earth" and these are operationalized as system conditions. The TNS framework supports processes in a structured way with the so-called ABCD process as a practical guide. First, **A**wareness and a shared understanding of sustainability, based on the system conditions, are created. This also leads to a general vision of how the organization or area would function if it were sustainable. Second, the stakeholders carry out a **B**aseline analysis to get a clear picture of the current situation and identify the challenges and opportunities at hand. By confronting the current situation with the general vision, numerous **C**reative solutions arise[4]. In the **D** step, an action plan is made in which priorities are set, by which smart early moves from the C list are chosen. Basic guidelines for this are: (1) let each investment be a flexible platform for coming investments that are likely to work further toward success as defined in your new vision and goals. In doing so, strike a good balance between (2) the direction and advancement speed with respect to the sustainability principles and (3) return on investment (to ensure sufficient influx of financial and other resources to sustain transitions).

---

[4]This is similar to the well-known GAP analyses.

Psychological and cultural factors are only implicitly addressed but at the same time TNS underlines the need to start with actors and stakeholders and their needs, motivations, and interests, instead of focusing on means or resources. TNS seems therefore suitable for creating awareness and articulating action goals (agency), less so for association.

The ABCD process can with attention, especially when basic human needs are emphasized, be used for creating association and passion. However, psychological and cultural barriers are not made explicit, and this process will be highly dependent on the skills of the facilitator. Visioning is an integral part of the process, because the four system conditions are very open, they leave the inspiration up to the stakeholders, which may lead to suboptimal results. The approach explicitly calls for leadership. The funnel metaphor shows a restorative perspective which gives hope for an even better future which is positive. It explicitly underlines the normative nature of the idea of sustainable development, for instance balanced material flows or equity.

A oneliner summary of TNS could be: Things get organized and structured

TNS case: Madison

Madison (200,000 residents) is the state capital of Wisconsin (USA). It is also a very green community in an environmentally conscious state. TNS was adopted by the city council. TNS is included in the city's Community Sustainability Plan (CSP), with concrete actions that are mostly focused around the four system conditions. This is primarily aimed at getting down to action. According to the city employees, the TNS approach is successful because it focuses on creating solutions, instead on just focusing on the problem. Next to the city's operations itself, it is used in the operations of many organizations that are closely tied to the city, like the Olbricht (botanical) Gardens, the conference center, and the fire department.

Both the employees of the city and of Sustain Dane stress that the effort at getting toward full implementation of the ABCD model is not yet successful. Especially the part of getting to a compelling vision is not (yet) part of the sustainability plan of Madison. This seems to be something most of the people involved in sustainability and working with TNS in practice are struggling with.

Looking at public reception, the citizens of Madison do notice that there are sustainable efforts going on. However, knowledge about the TNS framework or CSP is not widely spread. And because of the small number of people within Sustain Dane that are really trained in TNS, small personnel shifts might have a drastic effect on the Sustain Dane support of TNS. So, effort needs to be put into training professionals in the TNS approach.

From Madison, TNS has spread out to other communities as well, like Monona, a small town very near to Madison. In this town, it has been more of a community effort, sparked by citizens, and later adopted by the town. However, when the main town official involved in TNS implementation left, the momentum and energy was lost. They are trying to regain the power behind the initiative, but the group of people involved is quite small. It is quite hard to get the initiative up and running again.

Lack of public knowledge about the approach because it was successfully embedded in local policy/lack of inspiration and creation of passion that could generate sincere and broad support in the community.

2.4.3   *Cradle to cradle*

CtoC is very suitable for, and actually presupposes, a multi-stakeholder context (waste equals food). However, the concept is mainly a design philosophy for manufacturing products and buildings. Currently, an explicit process design is lacking, but for a "how to do it yourself" workbook (Van der Werf, 2009). Which is a good thing but not yet an integrated approach that addresses the three aspects of agency, awareness, and association. However, presenting the clear and appealing vision of CtoC is known to create a lot of inspiration and passion, especially valuable if awareness is already present. The vision also leads to awareness on what is possible but tends to ignore the structure of the issues and the limits of human agency. The approach explicitly calls for association between the various stakeholders and the collaboration with organizations such as Drift and Urgenda helps to provide additional resources and "toolkits". It evokes a lot of agency, which often swiftly wanes because the barriers in the system limit the possibilities for implementation. We are still living in a world that systematically favors cradle-to-grave solutions.

CtoC is currently evolving, at least in The Netherlands, into an approach suitable for development of regions and public organizations in various regions or cities, like Almere, Texel, Waddeneilanden, Venlo Region. Applicability in the multilevel context is possible, as is shown in the case study, but there are large difficulties in making this strategy operational. CtoC is broadening its scope and increasing the number of intervention instruments.

CtoC explicitly addresses social, ecological economical, and ethical issues. Spiritual issues are implicitly addressed (Hannover principles), whereas political–institutional aspects of sustainability seem to be out of its scope. The distinction of cycles in the technical and ecological domain which should be able to accommodate all of physicochemical and biological output of our society clearly defines how sustainable systems should function. Social and cultural criteria are limited to the concepts of equity and diversity. Economic criteria are limited to the notion of adding value and seem to be inspired by the ideas of a steady-state economy. CtoC does not offer a measurement tool but has a certification system which implies that criteria can be derived from the basic notions. Further operationalization should, in time, deliver a measurement tool. But it is of course also imaginable that integration of measurement and analytical tools like MFA, LCA, and EF is a way forward for CtoC.

At the same time they do not address explicitly the psychological and cultural factors that do shape and color a process of transition. We do not say that there is no sensitivity to the issues just that it is not explicitly mentioned as a factor of importance in the literature (McDonough, 1998; McDonough & Braungart, 2002a; 2002b; 2003).

CtoC is itself an example of leading the way but it does not contain the approaches for creating and supporting leadership. This approach is inherently and explicitly highly positive and tempting.

CtoC as opposed to TNS and EF starts with a solution instead of a problem and thereby is best suited to generate agency.

A oneliner summary of CtoC could be: Things start to move

2.4.3.1 *CtoC case: Almere*

CtoC was applied within the municipality of Almere for designing a sustainable strategy to expand the city to twice its size. Special attention was paid to district Almere-hout noord. A central idea was that independence with respect to energy, closed cycles, and integration with natural processes and structures will benefit the community in many ways including safety and environment and will yield an excellent living space (Sociaal Duurzame Wijk). The process was designed to be integrated, involving all stakeholders supplemented by a selection of out-of-the-box thinkers. CtoC was used as a guideline and a common set of principles to judge ideas. The above-mentioned CtoC principles were used, plus: anticipate evolution. Simply put, leave room for changes in the future.

It has, so far, lead to the construction of community houses, business centers, and a better infrastructure for bicycles. A masterplan was created with very high levels of ambition and inspiration. A success is that the plans for sustainability have been incorporated into formal contracts with the national government. This may serve as an example for other cities and regions. However, the integrated process was difficult for the municipalities and some other parties. The high levels of awareness and agency of the design phase did not carry over into the implementation phase.

This is by the interviewed attributed to the process design: Particularly, there was a lack of strategy and association, e.g. it did not include the operational teams.

2.4.4   *Ecological footprint*

The EF can be used in a multi-stakeholder context to assess what the contribution of each stakeholder to a collective footprint is. It can be used to compare various sectors on various levels. It provides information that informs strategy. The concept is primarily and mainly a measurement tool for quantifying impacts of the behavior of individuals and systems on ecosystem services. It does not provide insight in and an approach for addressing psychological and cultural contextual factors.

EF in itself does not address social or economic dimensions of sustainability. But it is implicated because of the focus on the ecological dimension on which the two other dimensions ultimately rest. EF seems to assume that if you show the inequity expressed as the amount of bioproductive resources and services, which our planet produces on a yearly basis, people will want to change because they see the need. However as an accounting and measurement tool, it can provide clear criteria for physical and biological aspects of sustainability. It is not the comprehensive and transparent planning tool as is often assumed, but it is the best we have yet in the area of resource accounting.

It is very suitable for creating awareness of the agenda, scale, urgency, and relevance of sustainability, but not for awareness of the structure of the issues and the stimulation of human agency. There is no process method for creating association, inspiration, and passion. It does promote a vision of more efficiency. Filling in a footprint calculation together and discussing it will create some shared understanding of the earth as a limited system impacted by what we do. This is limited to decreasing our footprint. However during the expert meeting ideas were generated for developing a beneficial footprint which may stimulate people. Overall, it the approach that is most science-based where it concerns measurement and monitoring. It makes it crystal clear that we are literally destroying since the 1970s the basis for our wealth.

EF is, therefore, mainly a wake-up call and a monitoring tool but does not generate much agency or association because it lacks a process design and promotes feelings of guilt and therewith denial or withdrawal instead of inspiration. The assessment-based approach is useful in situations where there exists already some agreement about the need and meaning of the idea of sustainable development. It can be used to elaborate those meanings, strengthen consensus, and/or formulate all kinds of criteria. No single indicator or ecological accounting approach can answer all questions related to sustainability. Multiple indicators will always be needed (Opschoor, 2000, quoted by Costanza, 2000) but at the policy-level footprinting can be highly informative and combined with LCA's and MFA's. There is already a serious body of knowledge to firmly ground organizational policies in physical and organizational reality (Huibregts, *et al.*, 2008; 2010).

A oneliner summary of EF could be: Things get measured

2.4.4.1   *EF case: Dutch municipalities*

The EF was applied within a project concerning eight Dutch municipalities: Bergen op Zoom, Den Bosch, Den Haag, Leidschendam, Nieuwegein, Pijnacker, Zoetermeer, and Wymbritseradiel. During the period 1999–2000 the footprint of each municipality was calculated.

All municipalities saw the project as an exploration and useful instrument for creating awareness and PR. However, none of the municipalities have taken up the footprint in their policies. The only municipality that conducted a series of measurements is The Hague.

The stakeholders attributed the disappointing result to the lack of consensus on what sustainability is, how it can be measured, and the fact that there are significant political and economic barriers, that impede attempts to do things in more sustainable ways. The polluter still gets rewarded in our current political-economic system. Necessary measures were perceived as unpopular. There is a strong belief that this will only change when people become more aware of urgency and inevitability of a sustainability transition, similar to the conclusion by Senge concerning the lack of deep awareness. Vision and leadership are called for but not actively promoted.

Now that we have highlighted the strengths and weaknesses of each approach it is time to see if and how they could reinforce each other.

2.5 COMBINING APPROACHES: DISCUSSION

We hypothesized that combining three existing approaches, TNS, CtoC, and EF, could create synergy and lead to better results than just working with one of the three. We found a large amount of support, not evidence, for this hypothesis in the literature, the expert meeting, and the analysis of TNS, CtoC, and EF. We ended each subsection of Section 2.4 with a one-liner to highlight the main strength of each approach. In order to uncover possible synergies, we formulated three questions that we will deal with subsequently.

The three questions were:

1. What is required for a sustainability transition in a local and regional cross-sector multi-stakeholder context?
2. Is the combination of the natural step, CtoC, and the EF sufficient to cover these requirements?
3. What sustainability and process approaches and concepts are needed in addition?

Concerning question one, we established that a sustainability transition must have a scope that is quite impressive, as it involves life in all its facets and thus includes physical-chemical, biological, sociocultural, economic, political–institutional, psychological, and spiritual aspects. Multilevel and multidisciplinary inspired theory and praxis needs to be developed further to be able to include such a huge number of requirements into one (change) program.

Using an extensive "checkmark list" that would cover all requirements as the basis for an integrated change management approach is too cumbersome, not feasible and misses the point entirely. A checklist provides focus at the project and programme management level and sensitizes actors with regard to the interconnectedness or interdependency of the indicators in the frame, but it is insufficient to support a push toward a transition.

We will, however, need a change management approach that takes such a "long list" into account. ICT can play various roles here, most likely in a dynamic decision support system. Contrary to older decision support systems that tend to prescribe the most optimal decision making paths, we need systems that help create awareness and stimulate innovation (effective) AND business continuity (efficiency); like the suggested support for CIO's in Chapter 7 (this volume) because you always build on what is already there.

What is also required is a set of unambiguous indicators, currently there are many sets of indicators each addressing specific aspects of sustainability, on the one hand we see increasingly sophisticated ecological, carbon, and waterfootprints, greenhouse gas protocols, etc. and within the field of ICT measures or standards such as PUE (power usage efficiency), ECR (energy consumption ratio), EnergyStar, OpenDCME (data center efficiency measure), etc. But when a new socially embedded problem like sustainability arises, indicators are not only used to monitor and evaluate but can be used politically by means of persuasion, advocacy, benchmarking, lobbying, influencing, etc. all play a role.

Combinations of eco-efficiency and eco-effective solutions work better: synergy is clearly present. It is relatively easy to start with eco-efficiency. Additionally, it leads to savings, less use of energy, and resources. If management reserves or reallocates part of the savings, then an efficiency strategy could provide the resources to develop and implement eco-effective measures or develop eco-effective products and systems. Efficiency strategies can bring energy and resource use of products and systems down to levels where eco-effective approaches become more feasible. It is more easy to produce enough sustainably generated energy when total consumption of energy has declined by a factor 3 or 4 or more. There are also motivational and social side effects: engaging in an eco-efficient project or program leads to opportunities to create a more deep awareness and association. And that is what is required to come to a transition at the sectoral level.

Considering the combination of the three approaches, question 2, we found that TNS could provide the process design lacking in CtoC and EF. But we also concluded that CtoC already understood this, they are extending their "toolkit" and associate with complementary partners. EF on the other hand provides a measurement tool that is internationally recognized and footprinting approaches provide the building blocks for a green accounting approach. It is also a fact that it is increasingly being recognized in policy circles and serves as a yard stick for companies.

The approach of CtoC complements TNS and EF, creating inspiration and passion and a firm belief that we can change the world into an even better place than it currently is.

TNS could provide the step-by-step strategy that would connect efficiency with effectiveness. When looking at vision as a requirement, the visioning process of TNS is very open and leaves room for the emergence of shared values that are reflected in the designs and strategies the view of CtoC could give more focus to the process and therewith. The two approaches seem to be complementary or even synergistic at a strategic level (Van der Pluijm *et al.*, 2008). We suggest that EF could be taken into the equation to see if we are really making ecologically sound progress.

Expressed in a more popular fashion when engaged in sustainability you aim for a paradigm shift: a radical new way of thinking and doing supported, and driven by a new set of values as well. At the same time you have to make do with your old ways of working and thinking to reach a transition or experiment with new ones. Hence the broad, encompassing, and value-laden focus of most sustainability approaches. In other words, values gives direction to the change effort, whose exact outcomes are uncertain, requirements deliver the "check-marks" to stay on course and synergies can deliver a "turbo-boost" to any change effort.

And this easily leads at the individual, organizational, and societal level to feelings of discomfort that, their turn, manifest itself as denial, disbelief, scepticism, voice, anger, exit, etc. As the case studies exemplify. To keep up the momentum to reach a desired transition, more is required. We will need to combine approaches and that is exactly the third research question that we aim to give an answer to.

Is a combination of these approaches sufficient to cover all requirements to, was the question posed and we have to conclude that the three approaches come a long way but are not enough. A common "blank" of all approaches is the human factor. While the chapter showed, at the same time, that change is deeply human and very much related to psychology, especially when problems have a high degree of complexity. One aspect of human behavior, in relation to sustainability that receives a lot of attention in the research community, is energy use of households. We will use this example to introduce relevant psychological mechanisms and show the salience of the subject.

Behaviors related to household energy conservation can be divided into two categories: efficiency and curtailment behaviors (Gardner and Stern, 1996). Efficiency behaviors are one-shot behaviors and entail the purchase of energy-efficient equipment, such as insulation. Curtailment behaviors involve repetitive efforts to reduce energy use, such as lowering thermostat settings. Energy-saving potential of efficiency behaviors is considered greater than that of curtailment behaviors (e.g. Gardner and Stern, 1996). For instance, households may save more energy by properly insulating their homes than by lowering thermostat settings. It should be noted, however, that buying and using energy-efficient appliances does not necessarily result in a reduction of the overall energy consumption when people use these appliances more often (the so-called rebound effect). In a European research program, BarEnergy, a third category of behaviors, is identified: Shift to more sustainable and renewable energy technologies. This strategy differs from efficiency and curtailment behaviors as it is not aimed at energy savings, but aims for a transfer to a different form of energy to contribute to a better environment. Curtailment and technology-shifting behavior is where the ICT industry can create value. Not unsurprisingly both behaviors resemble the concepts of efficiency (curtailment) and effectiveness (technology shifting) (Emmert *et al.*, 2010).

Hundreds of factors influence behavior, but what can we as sector contribute? Green and Kreuter (1999) developed the PRECEDE_PROCEED model to plan interventions in order to change individual behavior. They describe three general categories of factors that affect behavior:

1. Predisposing factors are especially internal antecedents; they motivate the behavior. Examples are awareness and knowledge, social norms, subjective norms, attitude, self-efficacy.
2. Enabling factors are the external antecedents to behavior, belonging to the situation or context. These are the factors that make possible or easier to conduct the new behavior. Example is legislation.
3. Reinforcing factors are factors which provide insight into the consequences of an action, thereby motivating people to change behavior. Examples are feedback and financial rewards.

Any intervention seeking to promote more sustainable human behavior should therefore contain interventions of different kinds:

1. Antecedent strategies: influence determinants beforehand, for instance by providing information about the savings that can occur when insulating your home. Another example: the increase in legitimacy of a business when it invests in the restoration and sustainable exploitation of shared water resources (and communicates about this). This strategy influences the actor. In the first example the consumer is influenced; in the second example a business is influenced. Other actors could be: policy-arenas, regulators, and citizens.
2. Consequence strategies: influence determinants after behavior is executed, for instance by providing feedback about last months' energy use at the same time the energy bill needs to be paid. Or by means of incentives like a discount when an actor uses less energy than expected. This strategy also strives to influence the actor directly. The first example is again a consumer example while the "discount" example also works for organizational actors like businesses.
3. Enabling and motivating strategies: satisfiers–dissatisfiers provide the means for an actor to actually perform the behavior or to desist from certain behaviors. In terms of energy, you can think about enabling local production both on the consumer and the infrastructure and transport side. An important subset of these strategies is: Removing barriers. Barriers can be real or perceived, they can be physical, political, legal, social, economic, or normative. The way in which Germany stimulated local generation of renewable energy serves as an example. The financial barriers to invest were removed and led to an explosion of installed power. Of course critique can be formulated about many details of the execution of this strategy but the general direction is clear: move to energy production that can be sustained over a long period and contributes to the competitive power of the German industry. Both strategies change things in the surrounding and influence the context of the actor.

2.6   CONSEQUENCES FOR THE ICT SECTOR

How does all this look when applied to the ICT sector? We have witnessed an increasing wealth of approaches, indicators, and frameworks being applied to sustainability goals in the ICT sector. We find that it is not necessary to strive for one encompassing framework for sustainability in the ICT sector, and that it is not needed (in fact it is undesirable) to have one indicator that indicates the level of sustainability reached.

Therefore, it is better to adopt approaches that reinforce each other. Which of the current approaches is best suited in a specific situation cannot be determined beforehand. However, some guidelines can be formulated:

- Ensure that the combination of approaches addresses the factors that are summarized in Table 2.1.
- The scope of the chosen approaches should be consistent with the scope of the company.

2.6.1   *Translation of TNS, CtoC, and EF to the ICT sector*

The studied sustainability approaches can be translated to indicators that are used in the ICT sector. The EF-like approaches (convert sustainability aspects to an appealing and quantitative measure) have gained most traction. TNS has been applied to the IT sector. The claim (Girshick *et al.*, 2002) is that "there is potential for IT to transform modern business into a more efficient, cyclical, networked, and sustainability-oriented system that pays returns to a "triple bottom line", through economic, ecological, and social prosperity". CtoC receives attention in terms of recycling and refurbishing old computer equipment. Another approach that might be seen as CtoC in software development is the re-use of existing components: when doing so, the footprint of the production stage of software becomes smaller. Software production involves energy, equipment, travel and other human activity that has its footprint. However, it can be questioned whether this will have a net positive effect: it has been demonstrated that careful reconstruction ("refactoring") of software can have a big effect on the energy needed to run that software[5].

2.6.2   *Scope of the approaches*

One feature of the ICT sector is that activities involve not only physical objects but also virtual objects. The creation and use of virtual objects has an impact as well. This might be overlooked by some traditional sustainability approaches, so when adapting frameworks from the non ICT world, it is good to keep in mind that:

- ICT equipment is produced, used, and decommissioned.
- ICT applications are produced and used.

Within ICT, most attention is paid to the energy usage part of sustainability. Especially during the usage phase, energy used during production receives less attention. Other "planet" aspects such as material scarcity of toxic materials are recognized as problems, but broadly accepted indicators for these aspects are not yet available. Other sustainability aspects of ICT, namely the "people" part, is yet to be developed (Bomhof and Hoorik, 2010).

Another outstanding feature of ICT is its transformative power: information may completely change the way a physical process is executed. It may even lead to the disappearing physical processes. An example is video meetings instead of physical meetings. Therefore, the rebound effect of ICT is quite huge. This is even stronger because of technological advances according to Moore's Law, that makes it easy to introduce ICT into every technological feature in the things that surround us.

A final remark on scope of indicators in the ICT sector is that indicators seem to be constructed in a bottom-up fashion. Evidently, when it comes to energy usage, it is most straightforward to assess the energy usage of individual components or at least at the level of identifiable ICT service units, such as data centers. When it comes to "greening" of the ICT sector, data centers are an easy target. Appealingly straightforward measures like the power usage efficiency are well established. However, sustainability of other levels in the ICT landscape should receive attention as well. For instance:

- Equipment level: energy used by data center equipment, the network, end user devices
- Process level: efficiency of the (business) processes that are supported by ICT
- Company level: sustainability of companies that make use of ICT
- Impact of ICT on regions/cities/countries.

A decision maker will choose an indicator level that fits the level of decisions that can be made; so each indicator has its own user group.

---

[5] See for instance: http://www.sig.eu/en/News_and_publications/Publications/1239/_Need_software_Avoid_mantras.html

2.6.3   *Implications*

As discussed above, choosing one indicator is not a panacea; it will not lead to the desired result. This is also corroborated in the Chapters (7 and 9) by Kalsheim and Plessius. An illustration is the PUE, which is defined as the ratio between total energy used and energy used by ICT equipment. This can lead to perverse results. An example: in an overdimensioned data center, where lots of cooling equipment is running but where only part of the floor is occupied by working IT equipment, it is easy to increase the power efficiency by just adding a bunch of old, unused computers. In that case, hardly extra cooling is needed, but these old computers do use power. This means that the ratio of "total energy used" versus "energy used for IT only" becomes more favorable. The main problem is that PUE is a ratio, that can be made "better" by not only decreasing the numinator, but also by increasing the denominator. Additionally, PUE only considers the data center and tries to optimize that unit only, where the business requirements, that are the main raison d'etre for the data center, are taken for granted. Their logic is not questioned.

Therefore, combining the efficiency measures like PUE with a sector-specific measure like "amount of energy used per service or production unit", will already "bracket" some of the efforts.

2.7   DISCUSSION: WAYS FORWARD

We do not claim that the three approaches are sufficient and fulfill all requirements; there always needs to be room for evolution, but they can profit from each other. We also emphasize that EF, CtoC, and TNS are not the only combination possible. A combination of industrial ecology, biomimicry and EF in a business setting would also be an imaginable option. Alternatively combining EF, CSR, and biomimicry would also deliver promising results. We showed that, when used expertly, elements of different approaches can reinforce each other. At the same time we conclude that the three emblematic approaches suffered from a lack of attention to behavioral and spiritual issues.

And exactly that, as Senge also concluded, is what keeps us from really pushing for a needed transition to create a more healthy but profitable ICT sector. Deep awareness of what is required for sustainability is lacking. This shows at the level of the industry by the strategy chosen to become sustainable. The focus on efficiency is helpful and certainly part of the movement but needs to be augmented. However, developing and implementing such strategies is difficult not only because legislative frameworks (context), vested interest and accepted ways of doing (carbon-based economy) and deeply ingrained perceptions and attitudes at the individual level impede the ability to transform collective needs into desired personal agency. We lack global association (we are not one planet but many competing nations and firms) and such issues remain major stumbling blocks for massive transitions. Besides looking outward and facing the current aggravating state of affairs (TNS/EF) or entering a vision of everlasting prosperity (CtoC) as starting point for sustainable actions, we conclude that introspection and a focus on personal growth are complementary activities that grow in importance when issues become more complex.

Next to the triple P (people, planet, profit) or triple E (equity, ecology, economy) we propose that there is a need for a triple A, we need to learn how to create Awareness, develop Association, and come into action by developing Agency.

REFERENCES

Ballard, D.: Using learning processes to promote change for sustainable development. *Action Research, Sage* 3:2 (2005), pp. 135–156.

Benyus, J.: *Biomimicry: innovation inspired by nature*. Perennial, New York, 2002.

Bomhof, F. & Hoorik, P. V.: Assessing the positive and negative impacts of ICT on people, planet and profit. *Proceedings First International Conference on Green ICT (IC GREEN)*, 2010.

Edwards, A.R.: *The sustainability revolution: portrait of a paradigm shift*. New Society Publishers, Canada, 2005.

Ehrenfeld, J.: Industrial ecology: a framework for product and process design. *Journal of Cleaner Production* 5:1 (1997), pp. 87–95.

Emmert, S., Lindt, M. & Luiten, H.: Barriers to changes in energy behaviour among end consumers and households. Bar Energy, final report, Oslo, 2010.

Gardner, G. & Stern, P.: *Environmental problems and human behavior*. Allyn & Bacon, Needham Heights, MA, US, 1996.

Girshick, S., Shah, R., Waage, S., Meydbray, O., Stanley-Jones, M. & Smith, T.: Information technology and sustainability: Enabling the future. The Natural Step Working Paper Series, San Francisco, CA, 2002.

Green, L. & Kreuter, M.: *Health promotion planning: an educational and ecological approach*. 3rd edn, Mayfield, Mountain View, CA, 1999.

Huijbregts, M.A., Hellweg, S., Frischknecht, R., Hungerbühler, K. & Hendriks, A.J.: Ecological footprint accounting in the life cycle assessments of products. *Ecological Economics* 64 (2008), pp. 198–807.

Huijbregts, M.A., Hellweg, S., Frischknecht, R., Hendriks, H.W., Hungerbühler, K. & Hendriks, A.J.: Cumulative energy demand as predictor for the environmental burden of commodity production. *Environmental Science & Technology* 44:6 (2010), pp. 2189–2196.

McDonough, W.: Essay: a centennial sermon: design, ecology, ethics, and the making of things. *Perspecta* 29 (1998), pp. 78–85.

McDonough, W. & Braungart, M.: *Cradle to cradle: remaking the way we make things*. North Point Press, New York, 2002a.

McDonough, W. & Braungart, M.: Design for the triple top line: new tools for sustainable commerce. *Corporate Environmental Strategy* 9:3 (2002b), pp. 251–258.

McDonough, W. & Braungart, M.: Towards a sustaining architecture for the 21st century: the promise of cradle to cradle design. *Industry and Environment* 26:2–3 (2003), pp. 13–16.

NLRO: Innoveren met ambitie: kansen voor agrosector, groene ruimte en vissector. NLRO-report 99/17, Den Haag, The Netherlands, 1999.

Robèrt, K.-H.: Tools and concepts for sustainable development, how do they relate to a general framework for sustainable development, and to each other? *Journal of Cleaner Production* 8 (2000), pp. 243–254.

Robèrt, K., Schmidt-Bleek, B., Aloisi de Larderel, J., Basile, G., Jansen, J., Kuehr, R., Price Thomas, P., Suzuki, M., Hawken, P. & Wackernagel, M.: Strategic sustainable development – selection, design and synergies of applied tools. *Journal of Cleaner Production* 10 (2002), pp. 197–214.

Rotmans, J., Kemp, R. & Asselt, M. V.: More evolution than revolution transition management in public policy. *Foresight* 3:1 (2001), pp. 15–31.

Senge, P., Linchtenstein, B., Kaeufer, K., Bradbury, H. & Carroll, J.: Collaborating for systemic change. *MIT Sloan Management Review* 48:2 (2007), pp. 44–53.

United Nations: Recycling, harvest of valuable components. United Nations University, Tokyo, Japan, not dated.

Van der Pluijm, F., Cuginotti, A. & Miller, K.: Design and decision making: back-casting using principles to implement cradle-to-cradle, *facilitating sustainable innovations: sustainable innovation as a tool for regional development*. The Cartesius Institute, and The Province of Fryslân, Leeuwarden, The Netherlands, 2008.

Van der Werf, M.: *Cradle to cradle in bedrijf*. Scriptum Books, Schiedam, The Netherlands, 2009.

Wackernagel, M. & Rees, W.: *Our ecological footprint: reducing human impact on the earth*. Vol. 9, New Society Publishers, Canada, 1998.

APPENDIX: CONDENSED FINDINGS EXPERTGROUP

Table 2.2.   The opinions on the most important issues in sustainability transitions.

Group	Awareness (4)	Cooperation, community feeling and purpose (association) (3)	Systemic change (agency?) (3)	Clear criteria/ science based (4)	Inspiring vision (3)
CtoC	0	0	33	0	67
EF 25	0	33	75	0	
TNS	75	100	33	25	33

Table 2.3.   The opinions on the most important barriers in transitions.

Group	Psychological and cultural factors (8)	Lack of consensus on a systematic science based learning process design (6)	System organisation
CtoC	25	17	25
EF	25	0	37.5
TNS	50	83	37.5

Table 2.4.   The opinions on what is required to overcome these barriers.

Group	Awareness (7)	Communication (association) (4)	Cooperation (association) (4)	Financial and tax system including the environment (agency) (6)	Innovation, leadership passion (agency) (3)
CtoC	15	25	25	0	33
EF	15	50	25	67	33
TNS	70	25	50	33	33

Table 2.5.   The opinions on what is necessary to create the movement necessary for the sustainability transition.

Group	Awareness (5)	Involvement cocreation (association)	Find/use synergy in cooperation (association) (5)	Inspiration, passion, action (agency) (6)	Positive approach: tempting (4)	Positive approach: rewarding (4)
CtoC	40	33	40	67	25	0
EF	40	0	20	0	50	75
TNS	20	67	40	33	25	25

CHAPTER 3

Green IT current developments—A strategic view on ICT changing the global warming trend

Anwar Osseyran

The rapid growth of carbon emissions is alarming and ICT is playing an important role in changing this trend. On the other hand, ICT usage is growing exponentially and the carbon footprint of the ICT sector itself has already surpassed in 2008 that of the aviation industry. This chapter, on current developments in Green IT, deals with greening the ICT sector itself, as with deployment of ICT to green other sectors. Greening IT requires greening the datacenter where most of the energy is being consumed. Virtualization and proliferation of cloud services will make computing and data access ubiquitous for the users and will help increase the utilization factor of ICT infrastructure, therefore increasing its energy efficiency. Green software and data life cycle management help decreasing the overall footprint of applications and avoiding over-sizing of hardware. ICT has been widely recognized as an enabler for a low-carbon future and greening by IT has become the motor of the rapidly growing green economy. Smart Grids is one of the most promising areas of greening by IT, as it enables optimization of electricity consumption, decentralization of energy production and deployment of renewable energy sources. ICT has also a major impact on decreasing the large carbon footprint of the transport sector by optimizing logistics, improving filling rates, reducing congestions and reducing the necessity of moving people and material. Smart buildings use ICT for improving insight in energy consumption and reduce its carbon footprint. ICT enables rethink building design and energy use and making buildings part of the energy solution instead of the problem. Greening the industry is a must as the world is since 1987 in overshoot in terms of raw material consumption. ICT can help in improving the efficiency of the industry, optimizing supply chains and raw material (re-)use and leading to customer participation as a prosumer of the industry. ICT is also widely deployed for dematerialization of content delivery and Big Data will help understanding customer behavior and assist in improving the adoption and deployment of sustainability measures. Finally, to avoid a rebound in energy consumption as a result of improving energy efficiency, a holistic approach in Green IT is highly recommended.

3.1 INTRODUCTION: GREEN IT AND SENSE OF URGENCY

Sustainability was defined by the WCED report "Our Common Future", published in 1987[1] as "to meet the needs of the present without compromising the ability of future generations to meet their own needs". The world's sustainable use of energy and raw material reached in 1987 the limit of planet capacity and since then it is in overshoot. Sustainable use calls for finding a balance between the three P's: People (social equity), Planet (environmental protection) and Profit (economic growth).

The growth of the ICT sector was largely related to the ability of ICT to support those three pillars of sustainability. ICT helped economic growth, improve efficiency of energy and raw material use and widen social access to wealth and prosperity. This growth has subsequently led to an exponential increase of energy use and carbon emission by the ICT sector itself. In

---

[1] http://www.un-documents.net/our-common-future.pdf

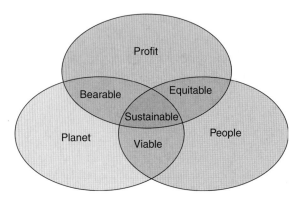

Figure 3.1.   The Venn diagram of sustainable development[2].

2007, Gartner estimated that ICT industry accounts for 2% of global CO_2 emissions[2], a figure equivalent to the emission of the aviation industry and despite the overall environmental value of ICT, Gartner believed this is unsustainable.

Surveys of the evolution of ICT consumption produce an alarming conclusion: those emissions grew from 2% of global emissions in 2007 to 3% in 2009 and are projected to increase to a colossal 6% by 2020, while globally the overall Green IT maturity is still low[3]. The promise of ICT is nevertheless that it can more than compensate the growth of its own emissions by greening other industries. The World Wide Fund For Nature (WWF)[4] and the American Council for an Energy-Efficient Economy (ACEEE)[5] quoted that "For every extra kilowatt-hour of electricity that has been demanded by ICT, the U.S. economy increased its overall energy savings by a factor of about 10". The much-quoted 'SMART 2020' Report[6] sets this 'offset factor' at a multiple of five. We are all assuming that Green IT will play that projected role of greening enabler, but the recent surveys of ICT footprint tell us that those expectations at the moment are not only exaggerated but even potentially fatal if that gap between potential and reality is not rapidly overcome with an accelerated multidisciplinary research, development and implementation program for both greening IT and greening by IT.

This chapter outlines recent developments in greening the ICT sector (greening IT) and in deploying ICT for greening other sectors (greening by IT). Two frameworks for Green IT will be first presented. Greening IT research portfolio outlined here is focused on greening the datacenter, as the majority of carbon emissions of ICT are produced there, improving software and data use of infrastructure and deployment of virtualization and cloud services. Greening by IT is being positioned as essential to resolve the present conflict between growth and sustainability. ICT will be positioned in this chapter as an enabler of the new low-carbon sustainable economy. ICT is key to green major economic sectors such as energy, transport, buildings, industry and media. A holistic approach for deployment and use of ICT is though necessary in order to avoid the rebound effect of overutilization of world resources due to the improved efficiencies brought up by ICT.

---

[2]http://www.gartner.com/it/page.jsp?id=503867
[3]http://www.ictliteracy.info/rf.pdf/green_IT_global_benchmark.pdf
[4]http://assets.panda.org/downloads/identifying_the_1st_billion_tonnes_ict_academic_report_wwf_ecofys.pdf
[5]http://www.aceee.org/sites/default/files/publications/researchreports/E081.pdf
[6]http://www.smart2020.org/_assets/files/02_Smart2020Report.pdf

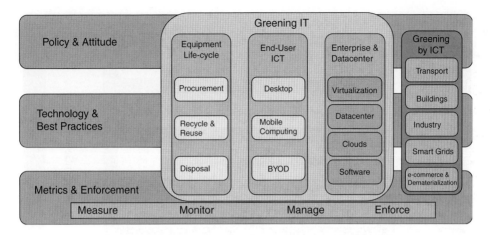

Figure 3.2. A variant framework of the one developed by Connection Research and RMIT University in 2010.

3.2 GREEN IT FRAMEWORK

One of the reasons for the mentioned gap between the high expectations and modest reality of Green IT impact on carbon emissions today is the lack of clarity in defining Green IT and understanding its many components. Green IT framework should cover both Greening IT and Greening by IT. The focus of Greening IT is on reducing the carbon footprint of the ICT sector itself. Greening IT requires collaboration between many parties including energy providers, installation equipment manufacturers, datacenter designers and service providers, IT equipment suppliers, software developers, universities & knowledge centers, designers and policy makers. Greening by IT is concerned with deployment of ICT for energy savings in other sectors. Greening by IT is per definition multidisciplinary and requires a close interaction between the ICT and the many application domains where ICT is a low-carbon enabler. All actions taken within a Green IT program should reflect policy, practice, attitude, technology, best practices, metrics and enforcement.

The framework presented above is a variant of that developed by Connection Research and RMIT University in 2010[7]. End user ICT is being fundamentally changed by the developments related to the present "bring-your-own-device"[8] (BYOD) possibilities in conjunction with public and private cloud offering. Many of the issues related to the use of ICT in offices will therefore shift towards the datacenters. As we will see later in this chapter, greening datacenters on the other hand, cannot be done in isolation without involvement of and collaboration with the energy provider. Datacenter energy label will depend on the ability to store hot and cold energy, and the possibility of becoming an intrinsic part of the smart energy grid. Greening by ICT must be conducted in close collaboration with and involvement of the ICT supplier as the success of the deployment of ICT will very much depend not only on the ICT applications developed but also on the willingness and ability of ICT vendors to conduct technological innovations in order to accommodate specific requirements in each area of industry. Community clouds and Big Data technologies will play an important role in maximizing the impact of ICT in greening other sectors.

---

[7] http://totalexec.posterous.com/latest-research-green-it-performance-internat
[8] http://www.gartner.com/resources/238100/238131/bring_your_own_device_new_op_238131.pdf

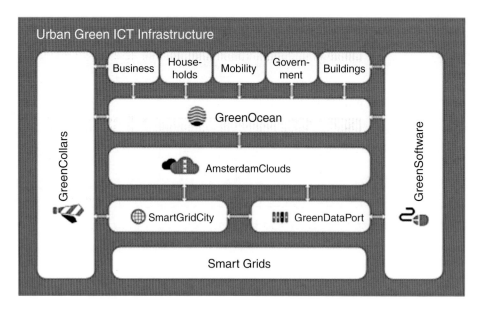

Figure 3.3.   Another variant framework adopted by the consortium green IT Amsterdam Region.

Another framework was adopted by the consortium Green IT Amsterdam Region[9]. Founded in June 2010, the consortium was initiated by the City of Amsterdam to help achieve a reduction of carbon emissions of 40% in 2025 as compared to 1990. The consortium members are regional datacenters, ICT suppliers, energy companies, regional authorities, and knowledge institutes. The framework adopted by the consortium is oriented towards meeting carbon reduction goals through a mix of projects making datacenters greener, electricity grids smarter, software more energy efficient, developing and incorporating sustainability skills within education and business and contributing to the creation of an ocean of green IT applications. The projects initiated within the framework are a mix of activities with quick wins and long term developments and investment.

3.3   GREENING THE DATACENTER: THE POWER LOSS CHAIN

One of the most important areas of Greening IT is greening datacenters. The British Computer Society published in 2008 a chart[10] stating the percentage of energy delivered (blue part) at every stage of the Power Loss Chain 'from fuel to CPU'. That chart gives a deep insight in the Green IT main issues that need to be tackled not only within the datacenter but also within the energy and ICT sectors as well.

The chart can be divided in three stages: The first stage, causing in absolute term the most significant losses, explains the importance of smart grids and power generation within the datacenter (as 65% of energy is lost at the power plant), the second stage in the middle of the power loss chain is concerned with greening the datacenter itself, and the third and last stage of the power loss chain, explains the high impact of virtualization and cloud technologies. Software and data life cycle management have an integral impact on the sizing of the equipment and total amount of energy needed. Finally, Big Data will provide better insight in user behavior, trends and correlations and will help improve energy efficiency of smart grids and ICT clouds.

---

[9]http://www.greenitamsterdam.nl/
[10]http://www.allhands.org.uk/2008/programme/download7004.html?id=1244&p=ppt

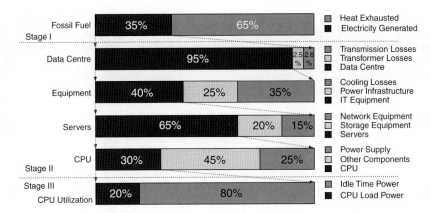

Figure 3.4.   The percentage of energy delivered (blue part) at every stage of the power loss chain "from fuel to CPU" as defined by Liam Newcombe (BCS, 2008)

3.4   GREEN IT AND SMART GRIDS (STAGE I)

Electrical smart grids enable optimization of electricity production and usage by producing, analyzing and using a wide range of information about the grid and its users. Smart grids are also essential in greening the power generation itself by offering citizens the possibility to generate sustainable energy like solar or wind energy empowering consumers to become a "prosumer" in energy[11]. Smart software is key in linking power usage data, analyzing patterns and trends, and enabling real-time balancing of energy production and distribution. Smart grids are therefore essential in the conversion of the power industry to become customer-centric and improve, however paradoxical this may sound, the efficiency of the power generation process by enabling decentralization of energy production. This will lead to the optimization of the power station size, minimizing its footprint and maximizing its utilization factor. Network and Big Data technologies will also play an important role in the transformation of electrical grid operations exploiting data sources in order to improve the production, power distribution and usage efficiency.

Smart grid technology is a rapidly evolving area of innovation and development[12]. Sensors and Smart meters are making significant changes in the utility industry offering the possibility of new applications and helping develop smarter buildings and smarter communities. Some of the biggest challenges in smart grids are currently the decentralization of energy production and the incorporation of new power storage possibilities like soil heat storage and electric vehicles as part of the smart grid. This requires new ICT technologies like demand-side management applications, and distributed generation management, including virtual power plants and local smart grids. The Science Park Amsterdam has already developed an R&D plan for experimenting in intelligent networks for smart grids with focus on applications and effects of the deployment of smart meters, decentral sustainable energy production and incorporation of the data center in the power grid[13]. Datacenters do have a large capability of conventional electricity generators that are rarely used and only deployed in the case of power interruptions. Those large capital investments with substantial CO_2 footprints could be better exploited if they form part of the smart grid itself and can be deployed for a better balancing of the power production at peek consumption periods.

---

[11] http://www.efficiencyconnectionsnw.com/uploads/01_Williams_1Dec2010_Macrowikinomics.pdf
[12] http://www.globalsmartgridfederation.org/documents/GSGFreport_stateofworldsmartgrid_4_26_12_000.pdf
[13] https://dl.dropbox.com/u/4042336/PowerMatcher%40Science%20Park%20Amsterdam.pdf

Research activities should provide a model for understanding how smart grids are evolving and analyze the impact of smart meters, regional power generation at homes or at the datacenter on electricity distribution, and customer behavior. Incorporation of energy production within the datacenter is a hot research domain and many innovative datacenter concepts are worth investigating[14,15,16]. Interesting new applications are virtual power plants, outage management and asset management. Interesting growth areas are smart grid software, data analytics and related applications and services.

3.5   GREENING DATACENTERS (STAGE II)

The global economic slowdown is forcing governments and organizations to reduce costs and avoid capital investment through consolidation of their ICT infrastructures[17]. The uncertainty about the future energy supply and rising costs of electricity are also putting the datacenter energy consumption in the spotlight. Enterprises are becoming more cautious about building new local datacenters, adding to the attraction of alternative models in the form of outsourcing, hosting and ultimately cloud computing services. Government rules and regulations to which datacenters must adhere do vary between countries and regions[18,19,20], however, datacenters are increasingly under the purview of regulatory and voluntary agreements on energy efficiency, green house gas (GHG) emissions, water consumption, and other environmental issues. Such moves increase the commercial benefits of green datacenter design and provide further momentum for research topics in, and for the market of, energy-efficient technologies.

Complexity of datacenters makes it impossible to have one single technology or design model to make a data center green. Greening an existing datacenter means a continuous program of improvement based on best practices and green technology guidelines. New datacenters must be green by design, built to meet the specific business requirements and use new technological possibilities. Higher computing power densities, frequently due to virtualization and higher dependency of businesses on ICT are putting additional strains on datacenter power and cooling requirements. Virtualization is however recognized as the innovation with the greatest impact on the shape of modern datacenters. It is also recognized as one of the most effective steps toward improving power efficiency of existing datacenters. In order to gain the maximum benefits from virtualization, the various components of the datacenter infrastructure must be optimized (research topic) to support more dynamic and higher-density computing environments.

The complexity of greening existing datacenter increases with the level of ambition. Low-hanging fruit is easy to pick, but higher ambitions require research into system innovations. There are many factors that play a role and the best results can only be obtained when greening is done on various fronts within the datacenter. The good news is that many solutions and techniques are available. Moreover, the greening process of existing datacenters cannot be forced because of the "installed base" or because of budgetary reasons and has to be done in steps to ensure continuity of the operations, minimize risks and consolidate gradual improvements. The strategy of discarding existing datacenters is far from green as it leads to high capital investment losses with a large already made carbon footprint.

However, improvements can only be truly assessed if both the energy consumption and datacenter productivity (and therefore the benefits the energy consumption provides) can be adequately

---

[14]http://bits.blogs.nytimes.com/2008/09/07/googles-search-goes-out-to-sea/

[15]http://www.ozzodata.com/

[16]http://www.parthenondatacentres.com/

[17]http://www.informationweek.com/news/government/policy/231900242

[18]http://www.datacenterpost.com/2010/10/government-regulations-impacting-data.html

[19]http://www.datacenterdynamics.com/focus/archive/2011/12/london-2011-eu-code-conduct-helps-telecity-comply-crc

[20]https://dl.dropbox.com/u/4042336/Eindrapport%20energiebesparing%20bij%20datacenters.pdf

measured[21]. Datacenter Power Usage Effectiveness (PUE) is a first, relatively simple step in achieving this goal. PUE hides yet almost as much as it discloses[22]. New datacenter productivity metrics must therefore be developed (research topic) in order to offer a better understanding of the datacenter's carbon footprint and make comparison between datacenter sustainability levels possible. Many research activities are conducted in this area[23,24,25,26].

Research on recipes and guidelines is being published to help with the transition to a greener datacenter[27]. An example is the development of a system that accurately measures the energy consumption of individual components (e.g. servers, routers, switches, network, storage, chillers, CRACs, PDU, UPS). The energy consumption on datacenter level, or on individual rack level, can then be continuously monitored, analyzed and optimized. Cooling plays an important role to lower energy consumption within a datacenter. The heat balance within a datacenter must be accurately mapped in order to avoid hot spots, hot zones or over-cooling. Computer simulation and modeling can help in this.

Datacenter operations are dynamic and require sophisticated management tools and a holistic view of the entire facility. It also requires a much closer relationship between facilities and IT professionals. SCADA systems help in monitoring the many factors that affect consumption and offer online reporting on live data making rapid intervention possible[28]. R&D in the area of datacenter management tools helps close the gap between cooling, power, and IT systems and offer a better insight into infrastructure and IT performance, enabling automatic adaptation to real-time changes in the environment. A comprehensive energy reduction plan can then be established and be regularly reviewed and updated. A tight coupling between the individual rack consumption and corresponding applications or users should be kept in order to support and stimulate the process of optimization. The ultimate challenge is to realize a closed-loop management and optimization of the datacenter energy consumption.

In addition to careful selection of energy efficient products, optimization of air flow and cooling (using measurements and simulations), and reducing the losses of energy transport and conversion (e.g. by locating the datacenter closer to the power source), there are additional measures that make a datacenter greener. Close research collaboration with the supplier can lead to the acceptance of higher operation temperatures than strictly specified in the operation manual. A few degrees higher operation temperature means less cooling power and a longer period of the free-air-cooling[29,30,31]. Collaboration with the energy supplier may help in getting excessive hot and cold energy stored or even used by third parties. And last but not least, the measures to green an existing datacenter are only optimal when the racks within the datacenter (or portions thereof) can be standardized. Modularization is therefore one of the key concepts in modern datacenter design. Most advanced form of modularization is the containerized solutions[32] with prefabricated components deployed on-site, and optimized for the applications served. A modular approach enables a standardized, application-optimized and tested model for the datacenter, offering a flexible approach to deployment. Modularization can also be seen as part of a shift towards an

---

[21] http://uptimeinstitute.org/wp_pdf/(TUI3009F)FourMetricsDefineDataCenter.pdf

[22] http://greenqloud.com/greenpowerusageeffectiveness-gpue/

[23] http://gigaom.com/cleantech/7-green-data-center-metrics-you-should-know/

[24] http://www.thegreengrid.org/~/media/TechForumPresentations2011/Data_Center_Efficiency_Metrics_2011.pdf?lang=en

[25] http://sameekhan.org/pub/W_K_2011_SUPE_SI.pdf

[26] http://link.springer.com/chapter/10.1007%2F978-3-642-32606-6_7

[27] https://dl.dropbox.com/u/4042336/CE_Delft_3686_Vergroenen_Datacenters_def.pdf

[28] http://www.sustainableplant.com/2011/04/data-center-monitoring-system-improves-efficiency/

[29] http://perspectives.mvdirona.com/2011/02/27/ExploringTheLimitsOfDatacenterTemprature.aspx

[30] http://www.datacenterknowledge.com/archives/2011/03/10/energy-efficiency-guide-data-center-temperature/

[31] http://www.eni.com/green-data-center/it_IT/static/pdf/ASHRAE_1.pdf

[32] http://www.datacenterjournal.com/press-release/welcome-to-the-worlds-first-data-center-container-park/

industrialized view of the datacenters. Standardization, modularization and virtualization are essential research topics for optimizing datacenter management, energy consumption, utilization factor and cooling strategy and do not stand in the way of innovation in ICT applications.

3.6   SUSTAINABILITY THROUGH VIRTUALIZATION AND CLOUD COMPUTING (STAGE III)

The move to cloud computing is one of the most dramatic ICT trends of this decade. The market for cloud computing services has continued to expand during the recession despite the decline in economic activity in most of the world. Cloud computing revenue is even expected to grow at a rate of 28.8%, with its market increasing from US$ 46.0 billion in 2009 to US$ 210.3 billion by 2015[33]. Cloud computing models – in all their diversity across public, private, community and hybrid clouds – will be the predominant paradigm for the next generation of ICT services. We are, however, only at the beginning as many cloud applications are still to be developed. New metrics and new levels of transparency are required if the impact of clouds on sustainability is to be adequately assessed.

Cloud computing is the next step beyond virtualization and grid computing. Increasing the utilization factor by sharing ICT infrastructure plays a pivotal role in greening ICT. Public cloud providers must make their datacenters as energy efficient as possible for obvious competitiveness reasons. New generations of hardware, software, and even business processes are developed in order to leverage the enormous scale of the cloud and make ICT more energy efficient. This promise can though be deceiving if the high utilization factor of mostly public clouds is not met. High availability requirements and projected economy of scale of public clouds may even play a detrimental role on energy efficiency below a certain utilization threshold[34,35].

As cloud computing continues growing, equipment suppliers and datacenter manufacturers will focus their designs on the needs of this important market segment. Energy efficiency is one of the most important among those. The green promise of clouds is in the massive investments in economy of scale and newest datacenter technologies, making clouds in general and public clouds in particular achieve industry-leading rates of efficiency, when and only when clouds are better utilized and less expensive to operate than traditional datacenters. This means higher levels of standardization, modularization, hardware optimization and virtualization. Public clouds are therefore most suitable for general purpose applications.

Community or sector-specific applications like in healthcare, chemistry or public services should be moved to community or sector clouds[36]. There is a rapidly growing need for a sector specific approach to cloud services. Public cloud services have so far focused on horizontal solutions. The business model of these services is based on maximum economy of scale, but is not always addressing sector-specific needs. In order to overcome this important limitation a sector-specific approach to cloud services is needed. Sector-specific clouds provide additional advantages over private and public clouds. They focus on those shared - and often specific - needs and facilitate business processes at the right level of privacy and security but also provide hardware effectiveness and efficiency within the sector. The implementation of sector-specific clouds will still be based on the general standard public cloud technology (open source or commercial) as underlying platform, but offers in the same time the opportunity to incorporate specific elements that are important for the sector. Common data and processing requirements can still be contracted within the public cloud, while enterprise critical data remain indoors or stored in the shared infrastructure. Sector-specific clouds are clearly complementary to the existing public and private

---

[33] http://www.pikeresearch.com/research/cloud-computing-energy-efficiency
[34] https://dl.dropbox.com/u/4042336/TNO%20-%20Cloud%20Computing%20-%20grijs%20of%20groen %202012.03.23.pdf
[35] http://greenmonk.net/2012/01/09/is-cloud-computing-green/
[36] http://arxiv.org/pdf/0907.2485.pdf

clouds but offer a much better fit between hardware configuration and sector specific application requirements leading to a much better application scaling than when using public clouds.

Many industrial sectors (health, creative industries, life sciences, logistics, etc.) still use physical media (disks or tapes) to exchange data, accepting all related disadvantages in terms of carbon footprint, time loss, higher costs and increased privacy, security and fraud risks. Sector clouds allow the pooling and integration of sector-specific solutions simplifying data and application sharing and creating a sector-specific cloud of users and providers with a sector-wide coverage. The ultimate goal is to develop an industry or sector specific ICT ecosystem. Another interesting advantage of sector clouds is that they can be set up in a federative way, making it possible to preserve the already made investments in local datacenter and ICT personnel. This aspect is mostly not addressed by public cloud providers, advocating a drastic outsourcing of ICT and ignoring the large scale destruction of existing private infrastructure and carbon footprint investments. Such a federative approach is still in the infancy stage and needs to be further developed (research topic).

3.7   GREEN SOFTWARE AND DATA LIFE CYCLE MANAGEMENT

The electronics industry has for years been aware of the energy challenge by the simple fact that the increasing miniaturization got limited by the cooling capacity within components and systems. The software industry must still make this turn. It is now often the case that software applications are written with Moore's Law still in mind. The direct software development costs are higher than CPU time and developers are no longer as in the eighties, interested in optimizing flops, bits and bytes. One even refers to the "Wintel" alliance: each new version of Microsoft Windows (and Office) required more CPU power. The users were consequently forced to upgrade their hardware, while they mostly only needed to buy a limited part of the functionality. Green labels for software are therefore important and research into best practices and tools for reaching that goal are very much needed. Green compilers should be developed to assess the level of energy efficiency of software developed or help with improving its carbon footprint. Tools should help the user select greener software components and commercial products for their applications. Software has a huge multiplier impact on datacenter energy consumption and developers know that a small effort of software optimization mostly leads to an order of magnitude of performance improvement, which means that much less hardware would be needed for the same performance level.

On macro-level, centralization of software will help decrease the overall footprint of applications and avoid over-sizing of hardware. The strategy of bring-your-own-device with hand held devices and thin clients or laptops, decrease the energy consumption of software and limit the footprint to the applications running in the cloud. Another interesting research area is how to decrease the data carbon footprint. The problem is the huge amount of unnecessary data duplication and transfer and the ease with which we do this to ourselves. The enormous impact on energy consumption can be reduced by using smart compression software. But the best solution is to adopt smart information lifecycle management (including the hot, cold and frozen data concept) and de-duplication techniques[37]. Smart software must ensure that data is not duplicated unnecessarily, kept online or stored for ever. Less unnecessary storage means less unnecessary energy waste.

3.8   BIG DATA AND SUSTAINABILITY

Big Data aims at exploiting advanced data storage, access and analytics technologies handling high volume and fast moving data in a variety of scenarios. These typically involve low signal-to-noise ratios, such as social media sentiment monitoring, or log file analysis, and require novel

---

[37] http://www.data-archive.ac.uk/media/2894/managingsharing.pdf

Big Data techniques in a user centric approach. The use of Big Data as a Green IT technology is already widely reported[38,39]. With smart phones and tablets being widely spread, apps are likely to be developed to offer the benefits big data applied in sustainability, in order to manage building energy efficiency, precision agriculture, food production, water, transportation and mobility, waste, manufacturing, and materials reuse.

As mentioned above, the energy consumption of data storage must be dealt with. Big Data would be adding a huge energy bill to the ICT world-wide consumption if this energy aspect of Big Data it is not adequately dealt with. Techniques for data management and de-duplication can help, but the biggest effect is expected to be obtained from Big Data technology itself. Big Data analytics are expected to give us a better insight in many socio-economic aspects that are difficult to obtain with traditional techniques of data analysis. Insight in user behavior, trends and correlations is very helpful not only in greening by IT but also in greening IT itself. Data analytics are expected to offer new insights in the use of ICT infrastructures and help improve energy efficiency of smart grids and ICT clouds.

3.9   SUSTAINABILITY AS MOTOR OF THE *NEW* ECONOMY

While most economic sectors in the world are shrinking as a result of the present global economic recession caused by the 2008 financial crisis, the so-called green economy is still growing. Transition to low carbon and energy efficient products and services offers a golden opportunity to escape the negative spiral the world is entangled in and helps deliver a sustainable economic system, with growing consumption but without destroying the climate and exhausting natural resources. Germany en Denmark's ability to resist the financial crisis has been impressive and their policies have ensured a strong position as low carbon pioneers and market leaders in renewable energy. 'Greening the economy isn't just good for the planet – it's good for the wallets, purses and pockets'[40] according to the British Deputy Prime Minister (February 2012).

The Davos 2012 Conference of the World Economic Forum, reiterated on the other hand that *water, food and energy security are chronic impediments to economic growth and social stability.* With more people, increasing consumption and fewer resources there is a growing global risk of severe tightening of water, food and energy resources to meet the demands of an increasing global population and may result in more wars and global unrest. And while ICT is part of the energy and pollution problem, ICT may be *the* most important part of the solution.

3.10   ICT AS AN ENABLER FOR THE LOW-CARBON ECONOMY

The above mentioned reports and many others[41,42,43,44,45] identify opportunities to deploy ICT to improve energy efficiency and dematerialize various goods and services. The SMART 2020 report was the first to quantify resulting carbon savings through ICT deployment. The report shows that whilst the ICT emissions are expected to increase as a result of more deployment of ICT (first order effect), this will lead to total emission reductions five times the size of the ICT

---

[38] http://techcrunch.com/2011/09/28/farmeron-google-analytics-for-farms-gets-investment-from-dave-mcclures-500-startups/

[39] http://gigaom.com/cleantech/using-big-data-to-make-solar-smarter/

[40] http://www.decc.gov.uk/en/content/cms/news/pn12_007/pn12_007.aspx

[41] ftp://ftp.cordis.europa.eu/pub/fp7/ict/docs/sustainable-growth/ict4ee-final-report_en.pdf

[42] http://e.mckinseyquarterly.com/W0RH00549D4B249DF830B29EECD040

[43] http://assets.panda.org/downloads/it_user_guide_a4.pdf

[44] http://www.oecd.org/site/stitff/45983022.pdf

[45] http://www.apt.int/sites/default/files/Upload-files/ASTAP/Rept-1-Introduction%20to%20Green%20ICT%20Activities.pdf

own footprint. The power of ICT is to change processes to become more energy-efficient (second order effect). According to Smart 2020, greening by ICT would enable up to 15% reduction of the global emissions by 2020 (business as usual, third order effect not included). The third order effect is that ICT teaches us about environmental effects of our behavior and enables us to rethink or change it. This third order effect, called transformative change by the WWF[46], typically an R&D area, is expected to have a factor three effect on the total reduction of global emissions (average of 45% reduction by 2020). ICT will help here to get beyond existing systems and transform linear thinking on infrastructure use, incentive models and even common values across societies and sectors.

Analysis of above mentioned Greening by IT initiatives does not provide a strict recipe how to deploy ICT to green the various sectors of the world economy. Most initiatives are just emerging from the pioneering phase and reveal many green business opportunities and regulatory issues. They provide also valuable insights on potential benefits (second order) of different IT applications. Third order effects are more difficult to capture, especially when socio-economic behaviors need to be predicted over a longer period of time. Various R&D attempts to systematically map the potential impact of ICT at a global scale show that most benefits can be obtained in the transport sector, smart buildings, smart industry, smart grids (dealt with in Part-I), e-commerce and dematerialization. The effects of greening those areas are best guesses and their priorities and potential impact will change substantially with technology and society.

3.11   TRANSPORT SECTOR

ICT has a major direct impact on greening the transport of people and goods in many ways. The role of ICT in optimizing transport logistics improving filling rates and transportation routes is well-known. Deploying sensing and control techniques, big data analytics, online management tools, GPS and mobile communications helps provide relevant information to the users, optimize traffic control, facilitate smart parking, improve public transport fill rate and quality of service, reduce traffic congestions and journey duration and implement dynamic road pricing scheme, minimizing as a result the total transport carbon footprint.

Another area of interest is minimizing employee home-work commutation by promoting work mobility (using BYOD[8], internet and cloud services) but also promoting concepts like smart work centers[47] (office centers in residential areas, shared by various employers and offering employees professional work environment close to their homes). Techniques like virtual presence (permanent video link with colleagues) help coping with many disadvantages of distant working like social isolation or degradation of team cohesion. Investments in next generation video conferencing techniques (High definition video, life-sized images, spatial audio, imperceptible latency and ease of use) help improve the quality of conferencing making external meetings less necessary and decreasing the company's travel carbon footprint. Saving commuting costs is also a hot item in education. Open and distance learning help to provide localized content to students in their native communities, save travel costs and increase availability of specialized knowledge.

A transformative impact of ICT is to rethink current transportation systems. Smart grids can lead to a new model where vehicles (electric cars with generators on board for instance) can be seen as mobile energy generators, where energy could be produced and used locally when not in use on the road. Such a shift could transform residential areas in people and goods transportation nodes. The same line of new thinking applies for datacenters, where putting datacenters in residential areas helps to minimize data transport costs and provides local energy production facilities as well.

---

[46]http://assets.panda.org/downloads/fossil2future_wwf_ict.pdf
[47]http://www.connectedurbandevelopment.org/connected_and_sustainable_work/smart_work_center

3.12   SMART BUILDINGS

ICT supports a wide range of technologies to improve efficiency of buildings and connect those to the internet and the clouds. Traditional technologies, such as Heating, Ventilation, and Air Conditioning (HVAC) are mature. Building Energy Management System (BEMS) and Building Information Modeling (BIM) are relatively new. Both are evolving rapidly and revolutionizing the usually conservative building sector. Proprietary BEMS solutions have already been around for several years but the trend is to push the concept one step further, offering an open platform for building management and modeling and for hosting users and third-party developers of application[48]. Building owners and occupants will then be able to tap data from their building system and other data sources (weather, energy providers, municipality, utilities, etc) and co-develop their own community building management system (prosumer approach). Social media communities help deployment of crowd sourcing for sharing best practices and funding the further development of smart buildings.

Smart controls have already been available for decades, but they have not yet been used in conjunction with a wide range of other data sources. With smart controls, smart buildings are able to offer efficient and natural lighting, better electrical appliances and cooking facilities and environmental heating and cooling systems. BEMS connected to internet and the clouds offer the users and operators a dynamic and precise insight into building energy consumption and assistance in saving energy. Cloud-based solutions connected to a wide range of real-time big data sources can offer the users, operators and managers recommendation for cost-effective energy efficiency measures to reduce costs and building carbon footprint.

A transformative impact of ICT on buildings is to make them part of the solution rather than part of the carbon reduction problem. Re-thinking buildings is already the trend for several years especially after the burst of the housing bubble at the end of the last decade. Energy self-sufficient houses are already available[49], and ICT plays herein a central role, with integrated BEMS solutions. Buildings will be designed with proper siting, architecture, material selection and making use of natural light, heating, cooling, and ventilation. Daylight will make it possible to use far less energy and even become net producers of electricity at little or no additional cost. Finally, deployment of high performance computing for better real-time simulations, combined with big data analytics should help both goals, rethink building design and use and dynamically manage energy resource production and consumption.

3.13   SMART INDUSTRY

The world trends in resource use and energy consumption show that current forms of industrial production are not sustainable in the long term[50]. Most heavy industries, like iron and steel, cement, aluminium, paper and pulp still follow a linear model based on extraction of natural resources, and disposal of products at the end of the lifecycle. The global footprint was 50% relative to the global bio-capacity and has doubled in 27 years. Since 1987 the world is in overshoot, currently 50%[51]. Many industries continue to be inefficient and wasteful and thereby over-consuming the limited resources of our planet. Industries need therefore to radically improve their energy efficiency, reduce resource consumption and avoid release of harmful by-products. ICT is crucial in reaching those 'must-do' goals. Industries which fail to become smarter will be less competitive and eventually lose their markets.

ICT contributes to the efficiency of industry at all stages, from the conception of products and services, to the design and control of plants, processes and production lines, to the choice

---

[48]http://www.pikeresearch.com/wordpress/wp-content/uploads/2012/05/SB10T-12-Pike-Research.pdf
[49]http://www.ieeeusa.org/communications/releases/2011/032211.asp
[50]www.footprintnetwork.org
[51]http://www.footprintnetwork.org/en/index.php/GFN/page/world_footprint/

of supply chain and marketing strategy. Use of IT-based controls and knowledge management systems within industrial production processes, in conjunction with access to real-time production and usage data, helps to improve operations, save energy and increase production efficiency. Software helps improving design-for-manufacturing and minimizing waste of raw material. A wealth of business opportunities is facilitated by ICT in the sophistication of product design and manufacturing processes, reducing material flows, cutting out waste in mining and manufacturing and dematerializing when possible. Dematerialization is gaining ground within the smart industries and the impact so far has been profound on the photo, music, film and paper industries.

Rethinking the current industry business models with ICT will migrate the industries closer to becoming service companies, focusing on what people want, rather than on mass-manufacturing products based on an ever-increasing demand for our planet's finite resources. ICT is indispensable in evolving the traditional industries into cradle-to-cradle service companies and producing stuff is a low-margin commodity business. Products will last longer— and with the support of ICT be recovered, reused, repaired, remanufactured, and recycled. This rethinking of the smart industry will create major new business opportunities with lower energy needs and better customer-oriented services. Customers buy light from Philips and not the light bulbs, and HP offers printing services instead of selling printers. Open ICT platforms for gathering customer information on products and services and involving customers in co-design[52], will make it possible to tap into the experiences, skills and ingenuity of millions of consumers around the world and involve them in the new generation of sustainable products (prosumer model).

3.14   ICT FOR DEMATERIALIZATION

The Telstra[53,54] and above mentioned GeSI Smart 2020[6] reports describe how ICT can help the global transformation towards dematerialization for a more energy-efficient and low-carbon economy. In 2012 GeSI, the Yankee Group and ACEEE published a more recent study[55] about the potential carbon reduction from broadband deployment for eight household activities: teleworking, e-news, e-banking, e-commerce, music and video streaming, e-learning, e-mail and digital photography. Although telecommuting (see also transport paragraph above) is responsible for the largest carbon reduction (85%) among the eight selected activities, the study shows that dematerialization plays a relatively growing role in carbon reduction of household consumption. Large energy savings will be made through the use of ICT for service delivery, substituting physical products with digital content (i.e. 'use bits instead of bricks'[56]).

On the other hand, waste management can be seen as a variant of dematerialization as it allows delivering products without the need for new raw materials. The race towards more processing power on one chip, characterized by Moore's law[57] leads to accelerated obsolescence of silicon systems and generation of rapidly growing e-waste. ICT should be deployed for the registration and recycling of electronic systems, avoiding the need to extract increasingly scarce raw materials, especially highly energy intensive such as rare earths. Recycling should ultimately lead to a sustainable consumption of natural resources and safe disposal of e-waste, reducing GHG emission and providing sustainability of supply to the industry.

---

[52]http://www.trendwatching.com/trends/customer-made.htm
[53]http://www.climaterisk.com.au/wp-content/uploads/2007/CR_Telstra_ClimateReport.pdf
[54]http://www.telstra.com.au/business-enterprise/download/document/business-industries-sustainability-executive-whitepaper.pdf
[55]http://www.telework.gov.au/__data/assets/pdf_file/0005/156668/Global_eSustainability_Initiative.pdf
[56]http://www.wwf.se/source.php/1183711/it_user_guide.pdf
[57]http://www.mooreslaw.org/

3.15   FINALLY, A WORD ABOUT THE NECESSITY OF AVOIDING RE-BOUND EFFECT

Are energy-saving light bulbs left on more than incandescent bulbs? Would the purchase of a fuel-efficient car lead us to use it more frequently than before? Examples of an often-overlooked rebound effect: when something costs less, we start to use it more. A report published by the Breakthrough institute[58] surveyed the literature on the rebound effect in energy efficiency measures and the implications effects for climate change mitigation policy. Multiple rebound effects seem to operate at various scales, having their greatest magnitude at the macroeconomic level with the potential to substantially erode much of the realized carbon reductions.

Literature shows that there is enough evidence of a rebound effect on an individual level. The Breakthrough report shows that 10–30% of energy savings from efficient cars and homes are lost by more use. Why this occurs is not well researched. Greater efficiency in buildings for instance doesn't frequently lead to the expected savings because of some unexpected socio-economic behavior of the tenants. But the highest rebound in energy use from green measures occurs in industry and commerce[58]. Improving the energy efficiency of production of raw material leads to lower prices, greater demand, more production, and consequently a rebound in the energy use as a compound result of the energy efficiency improvements. So, successful strategies for carbon reduction with ICT must include life cycle assessments in order to ensure that the new green system is really reducing carbon emissions. Another way of getting out of this vicious circle could be found in pushing people to consume less energy or, a more realistic approach, on using renewable energy sources so that it doesn't matter how much we consume. The lesson learned is that a systemic approach must be adopted when dealing with carbon reduction policy.

3.16   CONCLUSIONS

The world use of energy and raw material has been overshooting the limits of planet capacity since several decades and must be made sustainable in order to avoid threatening economic, social and environmental stability and deprivation of future generations. ICT is well positioned to help in this but a halt has to be put to the exponential growth of energy use by ICT itself. Besides putting a lot of effort into the development of energy-efficient ICT systems and equipment life cycle management, greening IT requires greening the datacenters, as the majority of carbon emissions of ICT are produced there. ICT can be made sustainable at three stages of the power loss chain: integrating the datacenter within the power grid, improving energy efficiency of the datacenter itself, improving software and data use of infrastructure and deployment of virtualization and cloud services. Research into smart grids and the development of innovative datacenter concepts as part of the energy ecosystem are worth pursuing, with interesting related areas like green software, sector clouds, Big Data analytics and related applications and services. ICT is essential to maintain growth and prosperity without conflicting with sustainability requirements. The focus should however shift from energy efficient linear processes from raw material to waste onto energy self-sufficient cyclic processes where raw materials are reused. Interesting research areas are ICT solutions for decentralization of energy production and deployment of sustainable energy sources, and the use of innovative ICT solutions for smart transport, smart buildings, smart industry and dematerialization. A systemic approach for deployment and use of ICT prevents a rebound effect of energy usage triggered by improved efficiencies brought up by ICT. This will lead to an energy efficient future where growth and prosperity will be possible without running the risk of a burst in an economic bubble caused by the exhaust of planet natural resources.

---

[58] http://thebreakthrough.org/blog/Energy_Emergence.pdf

CHAPTER 4

Higher-order sustainability impacts of information and communication technologies

Karel F. Mulder & Dirk-Jan Peet

Information and communication technologies (ICTs) are considered to be clean technologies. But underneath the surface of free flowing bits and bytes there are quite a number of issues which accelerate or oppose sustainable development (SD).

"Normal" innovation is aimed at increasing efficiency of products and processes. However, the ICT sector still has revolutionary characteristics: it creates new systems that provide new functions and services that do not merely make existing systems more efficient. Such new systems co-evolve with new types of behavior that might have both positive and negative sustainability effects.

These indirect effects of new ICTs are often more important than the direct effects and have bigger effects on SD. For example ICTs not only accelerate trends such as globalization of companies but also facilitate the easier and free spread of knowledge to educate people. They might not only diminish cultural diversity but also create opportunities for developing countries to access specific expertise.

In this chapter, we describe several indirect effects of the ICT sector that are relevant for SD. ICTs, like all other technologies, are far from being "neutral tools". ICTs will lead to different impacts in the industrialized world than in developing countries. This chapter will reflect on those impacts and the options to assess these impacts before introducing them.

4.1 INTRODUCTION

Over the past decades, ICTs have achieved a high speed of innovation. We have witnessed the emergence of a huge mobile applications industry in connection with the wide use of smart phones for professional and private purposes. It has changed our way of life and led to cultural changes, not only among youngsters and teenagers, but also among adults and professionals. Furthermore, computers, computer networks, electronic mail, and the Internet have resulted in new e-commercial industries and pervaded virtually every professional organization and our personal lives. Both IT and telecom have resulted in new infrastructures such as glass fiber networks and GSM transmission systems.

New products, services, and infrastructures do carry with them the promise that they are very beneficial in economic and social terms and have a low environmental impact, especially in comparison with traditional sectors and industries. However, the ICT industry requires a growing amount of energy, mainly in the form of electricity from the grid and from batteries. Data centers require megawatts of power and are often limited in their choice of location by the availability of grid capacity. E-commerce requires services of data centers and the fast handling, distribution, and delivery of small orders and may lead to a strong increase in transportation compared to traditional shop distribution.

The attentive reader might conclude that Jevons' paradox or the rebound effect is a second-order effect, which is absolutely true. These so-called second-order effects are very common, to give an example, energy use of data centers has been reduced dramatically, but the use of services has grown even more impressively, facilitated by the reduced energy costs. The second-order effect, growth of services, is in this example environmentally negative. Despite big strides in energy efficiency, the total growth outstrips the savings reached.

The challenge is to develop transition strategies that achieve positive second-order effects and contribute to eco-effective solutions, and as we have seen in Chapter 1 this does also include more efficiency. Before we can develop such strategies we need more insight on what second-order effects are particular to the ICT sector. This will be the subject of the remainder of this chapter and is also the subject of Chapter 5. In Section 4.2, we will look at how ICTs have developed over the last decades. The life cycles of ICTs are relatively short and user investments are relatively low, which gives different dynamics to the sector in comparison to classic sectors such as the base metal or the aircraft industry. Then we continue with a review of indirect effects and their contribution to SD. And we end the chapter with conclusions.

4.2   THE DYNAMICS OF INFORMATION AND COMMUNICATIONS TECHNOLOGIES

ICTs are an important part of recent social and economic transformations, both in industrialized as well as in developing countries. ICTs are considered to be the main drivers of the fifth longwave in economic development enabling decades of economic growth as described by Kondratiev (Freeman and Perez, 1988; Kondratiev, 1935). Kondratiev discovered that the state of the economy was not just determined by the normal market cycles of 8–11 years, but that there are also long-term cycles of about 40–50 years. These cycles are fuelled by new generic technologies that pervade and transform society (e.g. railroads, electricity, automobiles, and chemistry have created periods of economic progress. For example, the railroads created a transport infrastructure that allowed industry to settle at new locations).

The development of the ICT industry started at the end of the 1960s when prices of semiconductors decreased and chips and integrated circuits were increasingly applied in all sorts of devices. Over the last ten years various new applications have become available as handhelds. In addition, companies, governments, NGOs, and consumers have adopted ICTs as an indispensible part of their strategies and operations. Two important trends can be observed for ICTs:

1. The fast growth created new problems. There is now more attention for global environmental problems and poverty than 30 years ago.
2. ICTs became indispensable and more and more pervasive; without ICT most of our current affluence would evaporate.

The intersection of these two trends creates a number of complex issues ranging from increasing materials and energy consumption to the development of an unprecedented and hardly controllable global flow of information. In such undetermined situations, new technologies often lead to radical visions, both hopeful as well as visions of despair. By the end of the 1970s, for example, some people reacted quite negative to new information technology, as they expected massive unemployment and a loss of privacy (Busse, 1978). Others prophesied much more leisure time and "knowledge for all". It hardly makes sense to determine who is right: The real issue is how we can realize "hope" and how can we prevent the consequences that lead to visions of "despair".

For the ICT sector, it implies a more conscious determination of what kind of ICT our society needs, not just here and now, but also in the longer term, and what it implies elsewhere. This is a call to analyze the technological innovation process in a wider perspective, in order to serve the world better and create less risk. The way in which technologies are developed and used is partially determined by choices in the past. New technologies need to be adapted to existing technologies and infrastructures, and fit to the fashions of the day. This implies that technology develops in a more or less evolutionary way as an adaptation of an existing practice.

If several alternative technologies for the same function have been developed, then the relation between technology and its environment often pushes the development of a single standard. For example, around 1900 the internal combustion engine had to compete against electric vehicles and steam powered vehicles. After its victory, a transport infrastructure (including fuel supply) emerged that was completely adapted to the internal combustion engine. In turn this infrastructure made electric vehicles uncompetitive.

Such a historically determined process has consequences for future technology development as a new technology is generally based upon the existing standard. This process of technological evolution is called path dependency. As a result of dominant paths, alternative technologies are not feasible, i.e. the dominant technology has been "locked in". Adjacent technologies become dependent on the dominant standard and further contribute to "lock in" (Arthur, 1998; 1989). Moreover, users adapt their behavior to the availability of a technology, making it indispensable: A cell phone once was something optional, a more practical tool than fixed phones. Now the young people are asking the older generation "How did you live without it?" as they adapted their life styles to it.

The effect of lock in can also be observed in African telecommunication: The absence of an extended fixed telephone network created a stimulus for the transition to cell phones. Telecom operators were not bothered by strategic questions about investments in a fixed grid that still had to be re-earned. They could directly develop a market that developed almost in parallel to those of industrialized countries.

Of course lock in does not prohibit change. At some moment, the development of a technology can be regarded as a "dead end street", which might be an incentive for developing more radical alternatives. C-mos technology is often seen like that, and so research efforts have been devoted to developing radical alternatives like "photon computing" and even "quantum computing". However, the transition to such new technologies will only occur if such a technology is expected to match the existing C-mos technology in the short term. As C-mos has been optimized in many years of use, it is tough for new technologies to match the existing performance standards. The business case for private investments as alternatives for C-mos is therefore doubtful, which in turn becomes a self-fulfilling prophecy as this is hampering their development.

In the case of the software industry for example, Microsoft's Windows has become the standard operating system for personal computers. Compatibility issues reinforce the standard. Moreover, the belief in the standard, as being the standard, reinforces this standard too. Desktop PCs with a motherboard with CPU, hard drive, power source, fan for cooling, video adapter, etc. seem to be locked in as the standard configuration for office computing. This system is rather fixed and difficult to change. The reason for that is not just technological. Although it might be easy to equip every worker with a laptop and smart phone, few companies operate that way: A personal fixed workspace is seen as important. The office with a desk top and fixed phone line is more a result of organizational and cultural lock in than of a technological lock in.

On the other hand, we also observe the emergence of new technologies which interact in a flexible way with existing ICTs. This is an advantage of ICTs to more traditional technologies.

Development of new ICTs has specific characteristics that are especially relevant for SD. We review them and highlight the most important features that we will elaborate upon in the following section:

1. For software companies, the development of improved hardware is often crucial: improved hardware offers new options to software makers, persuading customers to buy new software with additional features. Such software cannot be used by older hardware. As computer users often need to use the same software as their partners, the "laggards" might be forced to buy the improved hardware as well to stay in the loop with their friends/colleagues. This cycle (improved hardware creates improved software creates a market for improved hardware creates more market for improved software) is a cycle that produces better computers and software, but it also creates more electronic waste. As such these effects are not intended by any individual actor; these result from the dynamics of ICT developments at a collective level.

2. The development of new *applications* (UMTS, software, digital services, etc.) and the ease of transmitting vast amounts of data result in a strong growth of data traffic. At the individual level, one might say that the sheer amount of data that every Internet user receives when online have transformed data into waste. The usefulness of various communications technologies has been seriously affected by spamming. At the collective level, a lot of energy is wasted to distribute information that receivers do not want.

3. The options to achieve vast market growth by developing innovative products and services induces the sector to lean heavily on existing approaches and techniques for its own physical support. Up until 5 years ago, little to no attention is, for example, paid to use the fast amounts of waste heat that are produced by large data centers. Using this heat is not easy, especially if it was not taken into account at the moment when the data center was constructed.
4. The ICT sector recently started to become interested in its indirect effects like:
   - Future depletion of some scarce raw materials;
   - the production of virgin materials with its environmental and social effects;
   - the production of semi-conductors and finished products;
   - the short economic lifecycle of ICT-applications leading to large amounts of waste;
   - the waste disposal and recycling processes and associated side effects on, for instance;
   - labor conditions in developing countries.

Given the characteristics of the dynamics in the ICT sector, the question is what indirect changes were caused by the introduction of new ICTs and what did these changes imply for SD?

4.2.1   *Higher-order effects of ICTs*

Technology is neither good, nor bad, nor is it neutral (Kranzberg, 1986)

The introduction of a new technology will always have an impact. This is the reason that a technology was introduced in the first place. The intended or foreseen effects of the introduction of a new technology are called first-order effects (Porter *et al.*, 1980). The first-order effect of, e.g. the introduction of text messaging was the ability of telecom engineers to communicate in an easy way among each other. Other first-order effects here were the use of batteries and devices for text messaging.

Second-order effects are effects caused by the options that the availability of a new technology offers for new behavior. In our example of text messaging, it became a novel way of communication and one of the cash cows of the telecom industry. This can be considered a second-order effect, which was not expected at the introduction of the technology. A third-order effect can be considered as the ban on cell phones in various situations where no information is allowed to be transmitted, e.g. exams; installations were developed to jam signals of cell phones in concert halls, in order to prevent the phones from interfering with the concert. In this case even a fourth-order effect can be detected because there are services available to remain connected despite jamming.

An example of another first-order effect is when databases of different governmental organizations are coupled. Potentially this leads to a better degree of control on the collection of taxes and the execution of regulations. But the balance of power between government and civilians also changes because civil servants now have more information at their disposal. This can tempt the legislator into refining the laws and complicating them. In turn, this can make the civilian feel powerless against the all-seeing, all-knowing governmental organizations. These are indirect or higher-order consequences.

4.2.2   *Second and higher-order effects on sustainability*

Environmental impacts are related to the size of the world population, the affluence of that population, and the resource efficiency of the technologies that are used to fulfill the needs of that population (Ehrlich and Holdren, 1971; 1972). Resource efficiency includes energy and materials efficiency, and also the efficient use of the limited scope that nature provides us to get rid of pollution.

In this chapter, the indirect impact of ICTs on SD is the focal point. In order to introduce a more efficient technology (which fulfills our needs with less sustainability impacts), it has to fit our culture (otherwise we do not want it) and the structure of society (otherwise it is illegal, or companies cannot achieve a profit). But society's culture and structure do not just determine the

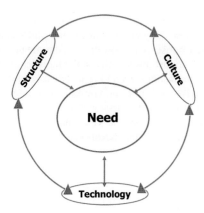

Figure 4.1. Co-evolution of society and technology.

acceptance of a new technology; society's culture and structure also are in turn affected by it. Society and technology co-evolve, see Figure 4.1.

Co-evolution does imply that you can distinguish clear causal mechanisms: The "rebound effect" is a well-known phenomenon. It is a typical second-order effect: New products that were meant to decrease the environmental impact increased it, because the increased efficiency of that technology led to a strong increase in demand. For instance since the 1960s, the power consumption of computation has diminished exponentially (in kWh per computation). It did not lead to a decrease in electricity demand for computing but facilitated an overall increase due to an even faster increase in the use of computers (Koomey *et al.*, 2011). This effect can also be seen in the strong growth of garden lighting due to the introduction of compact fluorescent lamps (CFLs) and LED-technology. The reduction of electricity costs of lighting a garden made it affordable for much more consumers. Hence, energy efficiency can result in a net increase of electricity consumption and its related emissions such as CO, CO_2, SO_x, NO_x, and soot into the atmosphere. This "rebound effect" (efficiency improvements boost consumption) is a special case of Jevons (1960) paradox. The same effect is also responsible for the strong growth of spamming, as "broadcasting" information has become so cheap. Rebound effects are mainly due to decreasing prices. But technologies might also create second-order effects by being used (or abused) in a different ways than was intended. For instance, the Internet created the infrastructure for various forms of crime, which in turn created a market for various forms of protective software. We will now discuss a number of second-order effects of new ICT's to give the reader a better feel where well-intended innovations aimed at more efficiency and reduction of energy use do not have the desired effects. In other words, thinking about the effects a technology exerts would help companies to design solutions that remain profitable over a longer period of time. Thinking about effects and safeguarding against undesired consequences is traditionally the role of governments and regulators but the recognition that we are all sailing in the same boat makes it more and more a mandatory part of any new development and innovation trajectory, especially if for instance, the precautionary principle becomes part of ISO standards. To see what effects might arise we give an overview of already well-known second-order effects in the ICT industry.

Job task sets
A second-order effect of the introduction of ICTs is that task sets of jobs have changed. Until the 1990s, secretaries were mainly used for diary and typing tasks. As word processors enabled office workers to take over these tasks, secretaries started doing other, more administrative and managerial types of work. In schools, lectures have sometimes been replaced by interactive software which changes the role of a teacher more or less into an educational coach and process facilitator.

On-line shopping

It is possible to buy virtually everything online starting initially with CDs, DVDs, books, games, holidays, but more recently also houses, cars, jewelry, art, etc. are bought online. This development holds a positive and a negative contribution to SD. If the logistics are handled in an efficient way, it can result in reduced energy consumption and reduced emissions, when compared to the case in which a consumer would drive to a book shop or a music store to buy the product.

On the other hand, the rebound effect is lurking. A large part of the products bought online is bought as additional consumption. As consumers can shop worldwide, there is a concentration of suppliers. Moreover, it has become clear that traditional down town shopping areas pay the price. This creates new problems for various cities.

4.2.2.1   *Dematerialized products*

ICTs can potentially cause savings in material use, dematerialization, by offering products in a digital way. For example music is bought online and newspapers can be read online. There is also potential for the paperless office in which much less paper will be used. However, the use of paper in industrialized countries has risen by 24% between 1988 and 1998 (O'Meara, 2000) due to the increased use of computers and the Internet in the office environment and the ease of printing. This shows the complications that arise when introducing a potential environmentally efficient innovation: Users do not change their habits and lifestyles just because the technology allows them to. The environmental gains are lost due to "devirtualization": Books and papers are still printed and music is burnt on CDs. However, it seems that a rebound effect also follows an evolutionary path because since the year 2000, a clear reduction of paper consumption can be observed. In The Netherlands, annual paper consumption per capita decreased from 233 to 171 kg (VPN, 2012). This is caused by less and thinner newspapers, the successful introduction of e-readers and indeed less paper consumption in offices (Banning, n.d.). It shows that societal adaptations that are enabled by ICTs take time, and there is no certainty that they will occur at all. For users, it takes time to adapt their habits and learn to handle new situations. But it also takes additional technologies, such as MP3 players, USB sticks, smart meters in households, etc., to enable users to fully utilize the environmental potential of an innovation.

Dematerialization of products makes it easy to acquire the product illegally. In fact, for many people, illegal copying does not feel like stealing as nothing is taken away from anybody. The sense of ownership is for many people related to "something material". At the same time, the ease of copying and the uniformity of the product obliged producers to sell their products worldwide at the same price. This created problems in developing countries as legal software was hardly affordable for any user. In a response to large scale illegal copying, Microsoft started selling cheaper versions of Windows operating systems, without some new features and options[1]. However, one should notice that piracy is not completely negative for Microsoft, as piracy also contributes to the "lock in", which gives Microsoft such a strong position.

ICT, globalization, and diversity

ICTs dematerialize information that used to be on a carrier. Naturally, this facilitates transport of the information. Designs can therefore easily be applied all over the world. This fluency of designs does not apply to the material products. One of the main effects of the dematerialization of information is that the world-wide variety in designs of products is decreasing: the same designs are just used all around the world (Mulder, 2006). A second order effect of ICTs is therefore a decreasing variation of products at the global scale. This does not imply a decrease of variation of products at a local scale. A consumer can still enjoy an unprecedented variation of products. However, this variation of products for sale is becoming more and more identical everywhere in

---

[1] For Windows pricing policies, see Wikipedia, for example for Windows 7: http://en.wikipedia.org/wiki/Windows_7_Starter

the world. In reaction, one can observe a positive re-evaluation of the local and regional-specific products, especially where food and services are concerned.

Cultural effects

In a world where designs are more and more international, the companies are increasingly international and the software used is predominantly available in English, there is a strong tendency to have all communication in English. Smaller language communities of the richer world are still of interest to software producers. But for many computer users in developing nations, knowledge of English is a precondition to operate a computer. Minority languages are under stronger pressure, especially if these language communities use their own set of characters.

So knowledge of English (or other widely spoken languages such as Spanish, Chinese, Arabic, Russian, or French), and therefore knowledge of the "Western" culture has become a precondition for being able to fully use computers and the Internet. This is a threat for local cultures; there is a wide availability of popular culture from "Western" sources, where the supply of popular culture from indigenous sources in indigenous language is virtually absent. On the up-side we are increasingly able to understand each other, opening up possibilities for integration, collaboration, and wealth creation, exchange and cross-fertilization of cultural memes (see also Chapter 10).

Global tele-working

A socio-economic effect of ICT and globalization is the outsourcing of back offices, software development and administrative tasks to developing countries. The main driver is reduction of (labor) costs. This is the freedom of trade, and one might argue that it supplies developing countries with jobs and the financial means for further development, i.e. wealth is shared. However, there are also accusations of exploitation and there is only a thin line between free trade and exploitation. An important point is the power in the contractual relationship between the "Western" firm and its tele-workers abroad. The effects of such teleworking are sometimes rather peculiar: Indian tele-workers operating on the North American market are taught to speak fluent English, with a regional accent in order to pretend that their call is local. They are even instructed about the weather conditions in the region that they will target for a day, in order for them to be able to present some small talk. But of course this can create moral conflicts as in effect, the tele-workers are paid to lie.

Local tele-working

An important and intended sustainability effect of new ICT systems was less commuting. Office workers were allowed to work (partially) at home by which overall traffic and congestion could be reduced. Initially, employers feared loss of control and productivity of their workers. Later, they also saw the benefits (less office and parking space required, increases in worker satisfaction). However, the number of people that started tele-working was rather disappointing. Although many office workers started tele-working partially, they were reluctant to work for the major part from home. A major factor in it is the feeling of being disconnected from the organization: not being present is a barrier for the informal communication of the organization (Daniels *et al.*, 2000) and increases the chances of not being seen or not being present when opportunities arise. Although, there are technical solutions that might help, organizations still need to learn how to deal with these issues: a main issue for the stagnant development of teleworking is mistrust—that the worker will not perform and that the employer (or competing colleagues) will not recognize the contribution someone is making, or that others tap into confidential conversations and e-mails.

Travel

ICTs were often regarded as offering substitutes for long-range travelling: you do not need to go somewhere in order to see it, you do not need to go to a person to communicate. ICTs can substitute travel. In the work place this holds to a limited extent. However, ICTs also allow developing new contacts that might lead to an increase in demand for travel. Clearly, world tourist travel did not suffer from the explosion of the Internet. "E-tourism" did not develop as a substitute for long-range travel, but now generally refers to online support tools for promoting tourism (Egger and Buhalis, 2008).

E-waste
Another indirect consequence of ICTs is e-waste. International exports of e-waste are regulated by the Basel convention. Determining what is waste and what is re-used is not very easy and often open for debate. Dumping of electronic waste in third world countries therefore still occurs. Large amounts of waste from ICT equipment are shipped to China which welcomes it as a source of cheap raw materials. Recycling industry sources estimate that between 50 and 80% of the e-waste collected for recycling in the western US is not recycled domestically but is exported to China (BAN/SVTC, 2000). Recycling in China often takes place under bad environmental and labor conditions.

4.2.3   *How to deal with higher-order sustainability impacts of ICTs?*

The standard reaction when people are confronted with higher-order impacts of new technologies is that these impacts cannot be foreseen,âŁ¦ and so we do not need to care. And that is a missed opportunity: a focus on second-order effects will lead to more sustainable forms of competitive advantage and the creation of market leadership. We can forecast certain second-order impacts: Jevons already showed in the nineteenth century that standard economic laws can predict the second-order effect of a more efficient technology. And of course you do not need any rocket science to predict that every piece of equipment once will be waste as long as we do not create far more eco-effective solutions.

It is in fact often clear that second-order effects of technology have really been foreseen: these effects were often dismissed as irrelevant by the proponents of a new technology. This points to a crucial issue: The introduction of a new technology is not a neutral activity, see also Chapter 2 on the use of indicators. It is driven by actors that have "put their guts in it to make the damn thing work" or more positively described, "it is a creative act of people with a creative urge" (Kidder, 1981).

Second-order impact of new technologies is often a matter of not willing to see these effects as they might endanger the whole enterprise that people have invested their brains and money in, very much a psychological thing. The issues discussed are often of a moral nature. Especially when third-order effects are predicted, arguments that point to disadvantages are easily marginalized as pure speculation. The sustainability advantages of new technologies are generally heralded, while opponents often tend to depict new technologies as environmental monstrosities (Smits, 2006).

New technologies lead to structural changes in society that involve moral values. What are marginal issues for one actor are key issues for another. Normally, there are established values in societies that we can refer to. However, when introducing new technologies or systems, new value specifications have to be developed for the new situations that the technology creates. The later a new consensus on value systems develops, the more problematic it might be. As we described in this chapter, technology often is "locked in" and change is becoming harder over time. It is therefore advantageous to aim at developing the moral values and adapting technological designs to those values, in early stages of technology development. Such a process is known as *constructive technology assessment*, and it aims at achieving a societal optimization of innovation. In short it tries to give substance to the people P, which is notoriously difficult as is also shown in Chapter 7.

4.3   CONCLUSIONS

In this chapter, the focus was on second- and higher-order (sustainability) impacts of innovations in the ICT sector. Developments in the sector are fast in comparison to more traditional industries. New versions of ICTs are introduced to the market at a considerable pace and ICTs become more and more pervasive. Next to that the demand for bandwidth increases accompanied by a growing number of services for ICT devices. These dynamics create a tendency to choose a proven technology where other factors and developments provide uncertainty. This may result in a faster lock in situations for relatively unsustainable technologies.

Examples of higher-order sustainability effects of ICTs have been described in this chapter. Main higher-order effects are related to dematerialization, (de-)centralization of expertise, and free flow of knowledge and data over the world. Acclaimed sustainability improvements take time for learning (for more information on how to speed up learning, see also Chapters 8 and 9). Learning processes should be organized in early stages of the innovation process because at that stage technology can more easily be adapted to the outcomes of the learning process. In this way the cost of failure of technological innovation can be reduced, potential volumes of sales increased, and contributions to SD more easily realized.

The main problem that prohibits a balanced assessment of sustainability effects of new ICTs in early stages of development is the fact that such innovation projects are driven by strong proponents that tend to neglect criticism. Discussing innovations in early stages of development with "outsiders" might help companies to take higher-order sustainability impacts of their innovations into account, and eventually adapt their innovations, contributing to SD as well as to the success of the company.

The second- and higher-order sustainability effects of ICTs are both positive and negative; they are certainly not neutral. However, if we act carefully in the early stages of an innovation we might strengthen the positive impacts while diminishing negative ones.

ICTs changed the world. The indirect sustainability impacts that were caused by ICTs are probably of a far greater magnitude than the direct environmental effects of the operations of ICT systems. Maximizing the ways in which ICTs contribute to sustainable societies requires strategies that support long-term value creation. Such strategies should be supported by long-term strategies of public authorities and one of the ways to influence innovations is by means of standardization, the subject of Chapter 5.

REFERENCES

Arthur, W.: Competing technologies, increasing returns and lock-in by historical events. *Economic Journal* 99 (1989), pp. 116–131.
Arthur, W.: Competing technologies: an overview. In: G. Dosi, C. Freeman, R. Nelson, G. Silverberg & L. Soete (eds): *Technical change and economic theory*. Pinter Publishers, London, UK, 1998.
Banning, C.: Duur papier is onze toekomst (expensive paper is our future). De Volkskrant, Amsterdam, The Netherlands, not dated.
BAN/SVTC: Exporting harm: the high-tech trashing of Asia. 2000, http://www.ban.org (accessed January 2013).
Busse, M.: *Arbeit ohne Arbeiter. Wem nützt der technologische Fortschritt?* (*Labour without laborers, who is taking advantage of technological progress?*). Fischer Taschenbuch Verlag, Frankfurt am Main, Germany, 1978.
Daniels, K., Lamond, D. & Standen, P.: *Managing telework*. Business Press, London, UK, 2000.
Egger, R. & Buhalis, D.: *eTourism – case studies*. Routledge, London/New York, 2008.
Ehrlich, P. & Holdren, J.: Impact of population growth. *Science* 171 (1971), pp. 1212–1217.
Ehrlich, P. & Holdren, J.: Impact of population growth. In: R. Riker (ed): *Population, resources and the environment*. U.S. Government Printing Office, Washington, DC, 1972, pp. 365–377.
Freeman, C. & Perez, C.: Structural crises of adjustment: business cycles and investment behaviour. In: G. Dosi, C. Freeman, R. Nelson, G. Silverberg & L.L. Soete (eds): *Technical change and economic theory*. Pinter Publishers, London, UK, 1988.
Jevons, W.: *The coal question, an inquiry concerning the progress of the nation and the probable exhaustion of our coal mines*. MacMillan & Co, London, UK, 1906, http://www.econlib.org/library/YPDBooks/Jevons/jvnCQCover.html (accessed February 2010).
Kidder, T.: *The soul of a new machine*. Little, Brown & Co., Boston, MA, 1981.
Kondratiev, N.D.: The long waves in economic life. *Review of Economic Statistics* 17 (1935), pp. 105–115.

Koomey, J., Berard, S., Sanchez, M. & Wong, H.: Implications of historical trends in the electrical efficiency of computing. *IEEE Annals of the History of Computing* 33:3 (2011), pp. 46–54.

Kranzberg, M.: Technology and history: "Kranzberg's laws". *Technology and Culture* 27:3 (1986), pp. 544–560.

Mulder, K.: Managing the dynamics of technology in modern day society. In: R.M. Verburg, J.R. Ortt & W.M. Dicke (eds): *Managing technology and innovation: an introduction*. Routledge, London/New York, 2006, pp. 109–129.

O'Meara, M.: Harnessing information technologies for the environmen. In: *State of the World 2000*. The Worldwatch Institute, Washington, DC, 2000, pp. 121–141.

Porter, A., Rossini, F., Carpenter, S. & Roper, A.: *A guidebook for technology assessment and impact analysis*. North Holland, New York/Oxford, 1980.

Smits, M.: Taming monsters: the cultural domestication of new technology. *Technology in Society* 28 (2006), pp. 489–504.

VPN: Historisch overzicht papierverbruik nederland (historic overview paper consumption in The Netherlands). 2012, http://www.vnp-online.nl/index.cfm?firm=vnp&fuseaction=show.page&pageid=152 (accessed January 2013).

CHAPTER 5

Standardization as ecodesign at sector level

Tineke Egyedi & Sachiko Muto

This chapter[1] analyzes standardization of mobile phone chargers to explore the role that compatibility standards might play in mitigating the negative impact of information and communication technology (ICT) on the environment. Building on insights gained from the economics of standards literature, we explore how the inherent effects of compatibility standards such as reducing variety, avoiding lock in, and building critical mass can have positive implications for the environment. We argue that current standardization literature and policy have overlooked this important (side) effect of compatibility standards.

Excessive diversity and incompatibilities in ICT generate e-waste, discourage re-use, and make recycling economically unviable; we therefore develop an economic-environmental framework for analyzing sustainability effects of compatibility standards and apply it to the case of mobile phone chargers. We conclude that well-targeted compatibility standardization can be equated to ecodesign at sector level and should be considered as an eco-effective strategy towards greening the IT industry.

5.1 INTRODUCTION

Following a much-publicized intervention by the European Commission, the mobile phone industry finally agreed in 2009 after dragging its feet for two years to introduce a standardized charger based on the micro-USB plug. In the Commission's communication surrounding the process, it was explicitly announced that by introducing compatibility there would be a reduction in the generation of e-waste and a significant benefit for the environment. The phone charger case illustrates that compatibility standards can contribute fundamentally towards improving the sustainability of the ICT sector.

In the remainder of the chapter, we first briefly consider the need to limit the direct impact of the ICT sector on the environment. (For a more elaborate discussion we refer to Chapter 1 and Egyedi and Muto (2012)). We continue with an introduction of the variety of sustainability-targeted standardization activities before looking more closely at the sustainable impact of compatibility standards as such. Based on their effects on the market, we extend our economics of standards framework to include implications for the environment. At the end of this chapter we discuss the potential use of compatibility standardization to achieve sustainability policy goals.

5.2 THE CHALLENGE OF SUSTAINABLE ICT

Implicit in many recent policy reports about the contribution of ICT as an enabler for sustainability in other sectors (Capgemini, 2009; Climate Group, 2008) is the assumption that ICT itself is a

---

[1]This chapter is based on "Interoperabiliteitstandaarden voor ICT: Een groene strategie in een grijze sector" (Egyedi and Muto, 2010, pp. 221–239). We revised it for publication in the International Journal of IT Standards & Standardization Research (Egyedi and Spirco, 2011) and again for the current volume.

clean sector. The negative externalities[2] generated by the sector are often disregarded. Indeed, in stark contrast with the immaterial notion conveyed by concepts such as "virtual", "web", and "the cloud", the impact of ICTs on the environment is highly concrete. It relates to the energy and materials used in manufacturing products, the packaging and logistics of distribution, the energy and material consumption during use, and the disposal at the end of a useful life. At each of these stages, standards can play a sustainability-enhancing role. Here, we focus on the two key problems of energy use and e-waste.

5.2.1   *Energy use*

ICT is responsible for a growing proportion of the global energy consumption and greenhouse gas emissions. In a high profile report titled "The Internet Begins with Coal" (Mills, 1999) already cautioned about the large amount of energy required for Internet use. He calculated that half a kilo of coal was needed to send a file of 2 MB. The energy consumption of the Internet, which was at the time 8% of the total energy consumption of the United States, was estimated to rise within 20 years to 30–50%. Although Mills was accused of exaggerating (Koomey *et al.*, n.d.), he was justified in highlighting the rapid growth of the Internet and the enormous amount of power which ICT requires (OECD, 2009a, p. 15).

In 2006, the electricity consumption of ICT in The Netherlands amounted to 8.4 TWh per year (Clevers and Verweij, 2007). This is equivalent to a capacity of 960 MW per hour. To indicate the extent of the problem, two nuclear power plants with the capacity of the one in Borssele, The Netherlands, would not be sufficient to generate this amount of energy. In this figure of 8.4 TWh the energy required for the mining of raw materials, the production of ICT, and the recycling of electronic devices are excluded. The Dutch ICT industry association ICT Office[3] estimates that the electricity consumption of ICT in The Netherlands is about 8% of the total Dutch energy consumption (ICT Milieu, 2009). In Europe, ICT products and services already consume 7.8% of the electricity and this proportion is projected to grow to 10.5% by 2020 (Forge *et al.*, 2009).

5.2.2   *E-waste*

The short lifespan of electronics also exacerbates the problem of their disposal at the end-of-life. Currently, global e-waste generation is growing by about 40 million tons a year (UNEP, 2009b) and is an increasing health hazard in the developing world (Williams *et al.*, 2008). Discarded ICT products are only partially repaired and re-used. While the bulk of used metals can be recovered from the remainder of the waste (Huijstee and de Haan, 2009, p. 8), other raw materials undergo changes during the manufacturing process. Chemical reactions with other substances makes their recovery impossible.

So far, the most ambitious attempt to address the e-waste problem has come from two EU directives. The European Directive on the restriction of the use of certain hazardous substances in electrical and electronic equipment (2002/95/EC) is designed to address the use of toxic substances in ICT equipment. The Guidelines on Waste Electrical and Electronic Equipment (2002/96/EC), WEEE in short, aims specifically at reducing ICT waste (Lieshout and Huygen, 2010). The WEEE is of interest because it makes ICT producers to a certain extent responsible for recycling their products, i.e. it aims at internalizing the environmental costs of the ICT industry.

---

[2] Externalities are the costs or benefits of a transaction incurred or received by members of society but not taken into account by the parties to the transaction. Externalities disappear when they are included in the cost estimate and become internalized. Externalities can be negative, e.g. polluting industry can drive down the value of houses in the area, or positive, e.g. a well-maintained park increases the value of houses in the neighborhood. (*Economics: An Introductory Analysis*, 1979). The environment-related negative externalities of ICT production and use are often not included in the calculated transaction costs.

[3] From January 1, 2013, the organization is know as *Nederland ICT*.

5.3 ECONOMICS OF STANDARDS

Scholarly interest in standardization has grown during the last two decades particularly from economists, which reflects Borraz' statement that "there is practically no economic activity nowadays that is not framed, whether partly or totally, by standards" (Borraz, 2007, p. 57). Unless indicated otherwise, in this chapter the term "standard" refers to committee standards[4]. A committee standard[5] is a documented specification "established by consensus, that provides, for common and repeated use, rules, guidelines or characteristics for activities or their results, aimed at the achievement of the optimum degree of order in a given context" (adapted from ISO/IEC, 2004, p. 8)[6]. This definition includes standards developed by formal standards bodies (e.g. International Organization for Standardization, ISO), standards consortia (e.g. World Wide Web Consortium, W3C), and professional organizations (e.g. Institute of Electrical and Electronics Engineers, IEEE).

From an economic perspective, committee standards represent three different functions (Egyedi and Blind, 2008, see the first column of Table 5.1). First, they have an informative function. They make life easier because we can refer to them and thus reduce informational

Table 5.1. Economic-environmental framework: Market and sustainability effects of compatibility standards.

Function of standards	Effects on the market	Effects on sustainability
Information	Increase market transparency Reduce transaction costs Correct adverse selection Facilitate trade	Ease consumer choice for green products and services
Compatibility	Create network externalities Increase competition Decrease vendor lock in	Counteract planned obsolescence and forced upgrades Extend useful life of products and their complements Facilitate growth of re-use markets and viability of green products
Variety reduction	Allow economies of scale Build critical mass	Reduce use of resources and waste Create favorable condition for investing in green innovations Ease automated disassembly and improve conditions for recycling Can increase environmental efficiency gains in process design

[4]The second sense in which the term "standard" is often used, is that of de facto standards, that is, specifications that underlie products and services with a significant market share. An example is the PDF specification of Acrobat Reader. Initially this specification was not meant to become a shared standard, that is, to be referred and built to by third parties; but its wide use has turned it into one.
[5]The term "open standard" is avoided because the debate it raises could distract the reader from the point we are trying to make. For the interested reader we point to Krechmer (Krechmer, 2006).
[6]We omitted "and approved by a recognized body" from the original definition in order to avoid a mostly theoretical—discussion about which standards body is "recognized" and which one is not, and to widen the term's applicability to widely used, nonformal committee standards as well.

transaction costs (Kindleberger, 1983). Such costs may entail, for example, the time and re-sources required to establish a common understanding. Standards reduce the costs of negotiations because "both parties to a deal mutually recognize what is being dealt in" (Kindleberger, 1983, p. 395). They reduce transaction costs between producers and costumers by improving recognition of technical characteristics and avoidance of buyer dissatisfaction (Reddy, 1990). They reduce the search costs of customers because there is less need to spend time and money evaluating products (Jones and Hudson, 1996). Because the information provided by standards increases market transparency, standards help to correct the occurrence of "adverse selection". Adverse selection takes place if the supplier of an inferior product gains market share through price competition because the supplier of a high quality product has no means to signal this information to potential consumers. Standards that contain information about quality will help suppliers to signal this information and minimize the likelihood that consumer selection is based on wrong assumptions. Moreover, because of increased market transparency, standards facilitate trade particularly in anonymous international markets, where parties to the transaction do not know each other.

Second, the committee standards we focus on also have a compatibility function. Compatibility can be of two types (David and Bunn, 1998, p. 172):

1. complementary when subsystems A and C can be used together, e.g. plug and socket; and
2. substitutive when subsystems A and B can each be used with a third component C to form a productive system, e.g. the USB interface of a digital camera (A) and of an external hard disk (B) vis-a-vis the USB interface of a laptop (C).

Technical compatibility is achieved by using gateway technologies, that is, "a means (a device or convention) for effectuating whatever technical connections between distinct production subsystems are required in order for them to be utilized in conjunction, within a larger integrated production system." (David and Bunn, 1998, p. 170). Notably, compatibility standards fall within this category. They interconnect and integrate subsystems in a way that allows subsystems from different suppliers to work together and replace each other. That is, standardized gateways loosen technical interdependencies between complementary and substitutive subsystems (Egyedi and Verwater-Lukszo, 2005), which is tantamount to increasing system flexibility (Egyedi and Spirco, 2011, p. 3), and reduce market interdependencies. They create a more open and competitive market. Because interfaces are standardized, consumers can more easily switch between providers and products, and are less easily locked in (Farrell and Saloner, 1985), enhancing the market's overall economic efficiency. Standards also create a more innovative market environment as they facilitate the emergence of standards-based clusters of new economic activity. Examples are the cluster of paper processing equipment and office products (e.g. printers, copiers, fax machines, binders) that has developed around the A-series of paper formats (ISO 216) and the cluster of Internet services based on TCP/IP.

The third main economic function of committee standards is that of variety reduction. It is heavily intertwined with the informative and compatibility functions discussed above. While the current formal definition of committee standard speaks more generally of "achieving an optimum degree of order in a given context" (ISO/IEC, 2004), an earlier definition of the Dutch standards body more explicitly mentions "[creating] order or unity in areas where diversity is needless or undesirable" (Beld, 1991). The principle aim of committee standards is to reduce needless and unhelpful variety by agreeing on a specification that can serve as a shared point of reference. Reduced variety mitigates economies of scale (i.e., cheaper units) and helps build the critical mass required for markets to take off. By reducing needless and unhelpful variety, the market becomes more transparent (information function of standards) and runs more efficiently (compatibility function). The first two columns of Table 5.1 summarize the main functions and market effects of compatibility standards.

The three economic functions of compatibility standards, that is, information, compatability, and variety reduction, offer a lens for identifying potential sustainability-enhancing effects of compatability standards (Section 5.4.2).

5.4 STANDARDS FOR SUSTAINABILITY

We start below with a brief overview of environment-oriented standards activities (Section 5.4.2) These differ from the compatability standards discussed in the subsequent section (Section 5.4.2) because they explicitly target sustainability issues. In the case of the compatability standards, increased sustainability is, as it were, a side effect – and a potentially significant one, as we will argue.

5.4.1 *Environmental standards*

There is a wide range of sustainability-oriented standards policies and initiatives. Some are general and thus also relevant to ICT. The 2010–2013 Action Plan for European Standardisation of the European Commission, for example, lists a number of proposed actions and mandated activities including the development of standards for energy efficiency (e.g. smart grids) and sustainability (e.g. air quality). Examples of generic standards initiatives are the management standards of the ISO 14000 and ISO 26000 series, and the standards for collecting data on, for example, climate change. ISO 14001 lays down basic requirements for the environmental management of an organization. Its model for continuous improvement is applicable to any organization interested in establishing and maintaining an environmental management system, including those in the domain of consumer electronics (Singhal, 2006). Examples of more specific standards in the 14000 series include ISO 14044 on the requirements and guidelines for a product life cycle assessment and ISO 14062 on integrating environmental aspects into product design. Likewise, ISO 26000 is a management system standard that provides organizations guidance on social responsibility. Its scope includes guidelines for corporate responsible behavior towards the environment, covering issues such as sustainable use of resources, climate change mitigation, and sustainable consumption.

Moreover, to evaluate and compare environmental policy options, agreed methods are needed for data collection as well as standard indicators and statistics. In this regard, the standards developed by ISO, IEEE, the Organization for Economic Co-operation and Development (OECD), and the World Meteorological Organization (WMO) play a fundamental role (e.g. ISO/NP 14067-2 on quantifying and communicating the carbon footprint of products).

Meeting the challenge of a sustainable ICT sector has also become an important focus for many standards organizations. The ISO/IEC Software and Systems engineering Sub-Committee (JTC1/SC27) has worked on life cycle assessments and developed a standard on the corporate governance of information technology (ISO/IEC 38500:2008). Recognizing that ICT is responsible for a growing proportion of the global energy consumption and greenhouse gas emissions, a large majority of standards initiatives focuses on the energy efficiency of ICTs during use. The Alliance for Telecommunications Industry Solutions (ATIS), the European Committee for Electrotechnical Standardisation (CENELEC), Energy Star, the European Telecommunications Standards Institute (ETSI), the International Electrotechnical Commission (IEC), ISO, and the International Telecommunication Union (ITU) have all developed standards aimed at improving the energy efficiency of electronic equipment. A number of recent initiatives are targeting the energy efficiency of networks and data centers (ITU, 2009a; OECD, 2009b).

5.4.2 *Sustainability effects of compatibility standards*

Most initiatives focus on the use of standards as a means to improve eco-efficiency. However, the ICT sector also needs to become more *eco-effective* (Appelman and Krishnan, 2010). Few people realize that eco-effectiveness can be a side effect of compatibility standards, and that in this sense compatibility standardization can be viewed as a form of ecodesign. We use the economic framework introduced earlier, that is, the main functions and market effects of compatibility standards, to explore this argument and induce in a systematic way the effects of compatibility standards as they relate to the environment. This approach allows us to formulate propositions that can usefully extend the framework to include environmental effects and serve as a starting point for further research. Table 5.1 column 3 summarizes the environmental effects mentioned in the propositions according to the most relevant – but in most cases closely interrelated—functions of compatibility standards at stake, i.e.: information, compatibility, and variety reduction.

Standards extend the useful life of products, peripherals, and accessories thereby counteracting planned obsolescence
By variety reduction and creating compatible complements and substitutes within and between (sub)systems, standards facilitate the replacement of some parts and the re-use of others. They facilitate refurbishing and replacement of spare parts and peripherals. For example, the carbon footprint of upgrading a PC with a new RAM module is significantly smaller than manufacturing a new PC. Compatible interfaces increase the possibility for such upgrades and therefore increase product longevity. The same reasoning applies to interfaces between products and their peripherals. Standardizing these interfaces extends the longevity of peripherals and accessories.

Standards reduce the use of scarce resources and e-waste
Because product parts and complementary peripherals are less likely to become obsolete when a system is changed or upgraded, fewer will be replaced (less use of resources) and discarded (less waste).

Standards counteract the environmental burden of forced upgrades
A side effect of compatibility standards is that standards loosen interdependencies between (sub)systems and their providers. They reduce the likelihood of supplier lock in and prevent situations in which consumers must keep abreast with forced ICT upgrades (e.g. because support for older software versions is withdrawn). Standards thus counteract the environmental burden of lock in and forced upgrades.

Compatibility standards facilitate the growth of re-use markets
Because standards facilitate refurbishing, replacement, and re-use, the value of second-hand ICTs increases. Without standardized interfaces, well-functioning printers, game consoles, etc. might not be in demand for re-use market because consumers cannot be certain that they will find compatible complements and substitutes (e.g. cartridges and games). In the past, re-use and re-manufacturing have not been considered important economic activities. They tend to be neglected in economic analyses (Williams *et al.*, 2008). However, the rising scarcity of resources makes it probable that these economic activities, along with recycling, will become increasingly significant.

A level playing field increases the viability of producing green ICT
Committee standards create a level playing field and increase competition among suppliers. Therefore, where there is demand for "green" products, companies will have an incentive to compete on sustainable product features. Consumer independence from vendors is a prerequisite to achieve such a market.

Standardization can help build a critical mass for products and thereby creates a more favorable condition for investing in green innovations
The uncertain outcome of wars between rival technologies can lead to a hold-up of investments (Williamson, 1979): producers will postpone investments for fear of investing in a "losing" system and having to write off sunk costs (i.e. costs that are specific and irreversible and which therefore cannot be retrieved). The same hesitations exist on the consumer side. Consumers will postpone their purchases and the market will stagnate. An agreed standard is needed for the market to take off. This also applies to the market for "green innovations" such as the need for

a standardized charger for electric vehicles. The European Commission has issued an industry-endorsed (ACEA, 2010) mandate to the European standards organizations for a common European charger on the grounds that incompatibilities would fragment the market and impede commercial success (European Commission, 2010).

A transparent market makes it easier for consumers to choose green ICT
Increased market transparency and reduced complexity due to information and variety reduction, and increased competition due to compatibility will make the ICT industry more susceptible to "green" consumer demands (e.g. energy efficient production and use of ICT, demands for repair and replacement of parts). A transparent market allows producers to charge more for sustainable products. Consumers who are aware of the negative environmental externalities of ICT will be better positioned to recognize and select "green" and upgradable products and services, and thus avoid adverse selection.

Standardized interfaces ease automated disassembly of products and thereby improve conditions for recycling
Ecological design is critical to green ICT because the energy and environmental costs of the product's lifecycle are determined in the design phase (European Union, 2009). Currently the emphasis has been on designing energy efficient ICT, which is also the main focus of the European Directive on Ecodesign (2009/125/EC). However, ecodesign can also extend to product design that has the end-of-life of products in mind, for example the design of ICT products with standard clips and fasteners to more easily disassemble and efficiently recycle them (Huijstee *et al.*, 2009). Standardized interfaces between complementary subsystems can facilitate the automated disassembly of products. Increased automation of such processes would make recycling less labor intense and heighten its economic viability.

Standards facilitate economies of scale which could, theoretically, lead to environmental efficiency gains
Compatibility standards ease larger-scale production. This could lead to environmental benefits if production processes are then organized in a way that is more efficient in terms of energy use and the raw material required. Similarly, efficiency gains in transport could also reduce energy consumption. However, facilitating economies of scale could equally lead to more production and associated rebound effects. This proposition should therefore be treated with caution.

The positive environmental side effects of compatibility standards are summarized in Table 5.1. They concern product design (e.g. replaceability of parts and recyclability of raw material) as well as process design (e.g. material and energy efficiency). In the terms of the theories discussed in this volume's introductory chapter, the main side effects coincide with three design principles of the crade-to-cradle approach (Braungart and McDonough, 2002: *Cradle to Cradle: Remaking the Way We Make Things*): elimination of the concept of waste; the output of a system should provide nutrients for the biosphere or the technosphere; and no downgrading of material should occur. In terms of the Life Principles in biomimicry (Benyus, 2002; *Biomimicry* 3.8)[7] compatibility standards foremost contribute to the principle "be resource (material and energy) efficient".

If correct – and elaborate research would be needed to further test, specify, and quantify these effects—the combined propositions would be a plea for considering compatibility standardization as specific form of ecodesign at the sector level. While such research falls outside the scope of this chapter, in the following we explore the initial tenability of our propositions in a preliminary case study.

---

[7] See http://biomimicry.net/about/biomimicry/lifes-principles, consulted 5 December 2012.

5.5   EXAMPLE: A STANDARDIZED CHARGER FOR MOBILE PHONES

In 2009 there were an estimated four billion mobile phones globally and about 500 million in Europe (Forge *et al.*, 2009). In 2008 roughly 1.2 billion mobile phones were sold with as many chargers. At least half of the sales was to replace "old" phones. A total of 51,000 tons of chargers were discarded that year (GSMA, 2009)[8].

5.5.1   *European Commission's initiative to standardize*

In February 2009, the European Commissioner for Enterprise and Industry, Günter Verheugen, made headlines, saying he was tired of waiting for industry to standardize the connector for mobile phone chargers and threatened mobile phone manufacturers with regulation. He was concerned about the amount of electronic waste produced by consumers throwing away incompatible but still functioning chargers when switching phones. Given that chargers can weigh twice as much as the handsets themselves, the effects of fewer of them being produced, distributed, and disposed would be substantial. It was not the first time the issue was raised. Already in December 2006 the Chinese government had announced that China would switch to the micro-USB connector (People Daily, 2006, December 19). The industry, represented by the Open Mobile Terminal Platform, responded to the Chinese call by announcing standardization on the micro-USB connector in September 2007. But for a long time nothing happened (ANEC, 2009).

Faced with Verheugen's ultimatum, the industry signed a Memorandum of Understanding (MoU) with the Commission[9] in June 2009 to develop a micro-USB standard. The European Standards Organisations (ESOs) were to facilitate implementation of the agreement. The standard was to lead to less hassle for consumers and have a positive environmental impact. In the words of the Commission: "The environmental benefits of harmonizing chargers are expected to be very important: reducing the number of chargers unnecessarily sold will reduce the associated generated electronic waste, which currently amounts to thousands of tons." (European Commission, 2009) While the MoU only covered EU territory, the market for mobile phones is essentially global. Therefore, the MoU was also expected to have an impact elsewhere, which was particularly relevant given the rapid growth of mobile phone ownership also in the developing world (UNEP, 2010). In October 2009 also, the International Telecommunication Union (ITU) announced that micro-USB would be its point of departure for work on the Universal Charger Solution. This "energy-efficient one-charger-fits-all new mobile phone solution" would result in a 50% reduction in standby energy consumption and be up to three times more energy efficient than an unrated charger (ITU, 2009b).

Although consumers were disappointed that the European Commission did not simultaneously address the connectors of other electronic devices (ANEC, 2009), Verheugen's proposal was widely welcomed—also by industry (UNEP, 2009a). It helped industry to meet the requirements of the European WEEE Directive[10].

5.5.2   *Environmental effects of a standard for mobile chargers*

The arguments the European Commission used to standardize mobile phone chargers illustrate many of the key environmental effects of compatibility standards discussed above. That is, a

---

[8]This is based on the estimation that 1.2 billion mobile phones were sold in 2008. Given that between 50% and 80% of these were replacement handsets, this would amount to 51,000–82,000 tons of replacement chargers every year (GSMA, 2009).

[9]The 14 companies who signed the MoU were Apple, Emblaze Mobile, Huawei Technologies, LGE, Motorola, NEC, Nokia, Qualcomm, Research in Motion, Samsung, Sony Ericsson, TCT Mobiles, Texas Instruments, and Atmel.

[10]For possible answers to the question why market coordination did not happen before see (Egyedi and Muto, 2010).

common standard will extend the useful life of chargers. It will facilitate their re-use. Fewer will be replaced, reducing the amount of e-waste generated. One (shared) charger will suffice for multiple phones. The storage of different types of plugs will be avoided—thus reducing the required amount of raw materials needed to produce additional plugs and reducing the amount of waste that would need to be managed. By standardizing plug and socket, a main cause of vendor lock in and planned obsolescence in the mobile phone market is addressed.

Our extended economic framework points to additional environmental effects as well, i.e. that such standards are also likely to:

- to facilitate the growth of re-use markets for chargers and mobile phones. Arguably, a second-hand mobile phone would have more value if one could be certain to find a charger for it.
- to provide an incentive to compete on and invest in sustainable product features by creating a level playing field (including design for automated disassembly and energy efficiency). Standardization is an important step in unbundling the compatible complements of phone and charger; and
- to make it easier for consumers to influence the market (i.e. demand and select, e.g. "green" chargers).

5.6   CONCLUSION: COMPATIBILITY STANDARDS—A GREEN STRATEGY IN A GRAY SECTOR

ICT is sometimes hailed too uncritically as a means to achieve sustainability in other sectors. The lack of sustainability of the sector itself is neglected. Predictions about the increasing use of ICT worldwide and the Internet, in particular, signal a growing, serious environmental problem. While there are major national and European directives on hazardous substances, electronic waste, and ecodesign, the perspective that compatibility standardization can have inherent positive environmental effects, as we argue in this chapter, has not been included so far as—at times significant—policy angle in its own right. While usually initiated for economic purposes, compatibility standardization has, as we argue, the potential to become a green strategy, contributing towards addressing the growing problem of energy use and scarcity of precious resources in the ICT sector. We recommend that this vantage point is investigated further.

Research is needed that examines whether compatibility standardization can, indeed, be regarded as a form of ecodesign at sector level. This requires, first, case studies that determine whether the environmental effects we identify are correct, complete, and/or have unforeseen rebound effects. They will help to further elaborate the economic-environmental framework developed in this chapter. Second, complementary quantitative data is needed about the environmental impact of compatibility. Third, although all parties, industry included, say they benefit from standardization, the example of the mobile phone chargers emphasizes that self-coordination by industry is not self-evident. We recommend a systematic analysis of the circumstances under which instances of market failure can take place in the light of our economic-environmental framework.

ACKNOWLEDGMENTS

The authors thank the three anonymous reviewers of the "Standardization and Innovation in IT" conference 2011, the reviewers of the International Journal of IT Standards & Standardization Research, and the editors of the current volume for their feedback and suggestions.

REFERENCES

ACEA: Auto manufacturers agree on specifications to connect electrically chargeable vehicles to the electricity grid. European Automobile Manufacturers' Association, press release, 2010, http://www.acea.be/index.php/news/news_detail/auto_manufacturers_agree_on_specifications_to_connect_electrically_chargeab (accessed January 2013).

ANEC: ANEC newsletter, no. 102. 2009, http://www.anec.org/anec.asp?rd=453&ref=02-01.01-01& lang=en&ID= 251 (accessed January 2013).

Appelman, J. & Krishnan, S.: Energy-efficiency and effectiveness: supplementary strategies for the IT-sector? *Proceedings IC-Green Conference*, Athens, 2010.

Beld, J. v.d.: Technische normen niet altijd commercieel gewenst. *Elektrotechniek-Elektronica* 2 (1991), pp. 22–24.

Benyus, J.: *Biomimicry: innovation inspired by nature*. Perennial, New York, 2002.

Borraz, O.: Governing standards: the rise of standardization processes in France and in the EU. *Governance* 20:1 (2007), pp. 57–84.

Braungart, M. & McDonough, M.: *Cradle to Cradle: Remaking the way we make things*. North Point Press, New York, 2002.

Capgemini: Trends in mobiliteit. Cap Gemini/Transumo, Utrecht, The Netherlands, 2009.

Clevers, S. & Verweij, R.: ICT stroomt door. Inventariserend onderzoek naar het elektriciteitsverbruik van de ICT-sector & ICT-apparatuur. Ministerie van Economische Zaken, Den Haag, The Netherlands, 2007.

Climate Group: Smart 2020: Enabling the low carbon economy in the information age. 2008, http://www.smart2020.org/_assets/files/02_Smart2020Report.pdf

David, P. & Bunn, J.: The economics of gateway technologies and network evolution: lessons from electricity supply history. *Information Economics and Policy* 3 (1998), pp. 165–202.

Egyedi, T.M. & Blind, K.: Introduction. In: M. Egyedi & K. Blind (eds): *The dynamics of standards*. Edward Elgar, Cheltenham, UK, 2008, pp. 1–12.

Egyedi, T. & Muto, S.: Interoperabiliteitstandaarden voor ICT: een groene strategie in een grijze sector. In: V. Frissen & M. Slot (eds): *Jaarboek ICT en Samenleving 2010*, 7de editie: *De duurzame informatiesamenleving*. Media Update, Gorredijk, The Netherlands, 2010, pp. 221–239.

Egyedi, T. & Muto, S.: Standards for ICT a green strategy in a grey sector. *International Journal of IT Standards & Standardization Research* 10:1 (2012), pp. 35–48.

Egyedi, T. & Spirco, J.: Standards in transitions: catalyzing infrastructure change. *Futures* 43 (2011), pp. 947–960.

Egyedi, T. & Verwater-Lukszo, Z.: Which standards characteristics increase system flexibility? Comparing ICT and batch processing infrastructures. *Technology in Society* 27 (2005), pp. 347–362.

European Commission: Harmonisation of a charging capability of common charger for mobile phones frequently asked questions. Press release, 2009, http://europa.eu/rapid/pressReleasesAction.do?reference= MEMO/09/301 (accessed January 2013).

European Commission: Towards a European common charger for electric vehicles. Press release, 2010, http://europa.eu/rapid/pressReleasesAction.do?reference=IP/10/857&format=HTML&aged=0&language=EN &guiLanguage=en (accessed January 2013).

European Union: Richtlijn betreffende de totstandbrenging van een kader voor het vaststellen van eisen inzake ecologisch ontwerp voor energiegerelateerde producten. 2009/125/eg, Europees Parlement en de Raad, Brussel, Belgium, 2009.

Farrell, J. & Saloner, G.: Standardization, compatibility, and innovation. *The RAND Journal of Economics* 16:1 (1985), pp. 70–83.

Forge, S., Blackman, C., Bohlin, E. & Cave, M.: A green knowledge society: an ICT policy agenda to 2015 for Europes future knowledge society. A report for the Ministry of Enterprise, Energy and Communications, Government Offices of Sweden, 2009.

GSMA: GSM World agreement on mobile phone standard charger. Press release, 2009, http://www.gsmworld.com/newsroom/press-releases/2009/2548.htm (accessed April 2009).

Huijstee, M. v. & de Haan, E.: E-Waste. SOMO, Amsterdam, The Netherlands, 2009, http://somo.nl/publications-nl/Publication_3289- nl/at_download/fullfile (accessed April 2009).

Huijstee, M. v., de Haan, E., Poyhonen, P., Heydenreich, C. & Riddselius, C.: Fair phones: it's your call: how European mobile network operators can improve responsibility for their supply chain. SOMO, Amsterdam, The Netherlands, 2009.

ICT Milieu: ICT Milieumonitor. ICT Office/ICT Milieu, Woerden, The Netherlands, 2009.

ITU: ITU-T Focus Group on CT and Climate Change, Deliverable 1: Definitions. ITU, Geneve, Switzerland, 2009a.

ITU: Universal phone charger standard approved one-size-fits-all solution will dramatically cut waste and GHG emissions. Press release, ITU, Geneve, Switzerland, 2009b.

Jones, P. & Hudson, J.: Standardization and the cost of assessing quality. *European Journal of Political Economy* 12 (1996), pp. 355–361.

Kindleberger, C.: Standards as public, collective and private goods. *Kyklos* 36 (1983), pp. 377–396.

Koomey, J., Kawamoto, K., Nordman, B., Piette, M. & Brown, R.: Initial comments on the internet begins with coal. Memorandum (LBNL-44698), Berkeley Lab, Berkeley, CA, not dated.

Krechmer, K.: Open standards requirements. *The International Journal of IT Standards and Standardization Research* 4:1 (2006), pp. 43–61.

Lieshout, M. v. & Huygen, A.: ICT en het milieu Mag het een bitje meer? In: V. Frissen and M. Slot (eds): *Jaarboek ICT en Samenleving 2010*, 7de editie: *De duurzame informatiesamenleving*. Media update, Gorredijk, The Netherlands, 2010, pp. 139–158.

Lipsey, R.G., & Steiner, P.O. *Economics: an introductory analysis*. Addison Wesley, New York, 1979.

Mills, M.: The Internet begins with coal: a preliminary exploration of the impact of the internet on electricity consumption. The Greening Earth Society, Arlington, US, 1999.

OECD: OECD communications outlook 2009. OECD, Paris, France, 2009a.

OECD: Towards green ICT strategies: assessing policies and programmes on ICT and the environment. OECD, Paris, France, 2009b.

People Daily: China spells out national standard for cell phone chargers. 2006, http://english.peopledaily.com.cn/200612/19/eng20061219_334047.html (accessed April 2009).

Reddy, N.: Product of self-regulation: a paradox of technology policy. *Technological Forecasting and Social Change* 38 (1990), pp. 43–63.

Singhal, P.: Integrated product policy pilot on mobile phones, stage III final report: Evaluation of options to improve the life-cycle environmental performance of mobile phones. Nokia Corporation, Espoo, Finland, 2006.

UNEP: Guideline on the awareness raising-design considerations: mobile phone partnership initiative project 4.1. United Nations Environment Programme, Basel, Switzerland, 2009a.

UNEP: Recycling from e-waste to resources. United Nations Environment Programme, Nairobi, Kenya, 2009b.

UNEP: UNEP yearbook 2010: New science and developments in our changing environment. United Nations Environment Programme, Nairobi, Kenya, 2010.

Williams, E., Kahhat, R., Allenby, B., Kavazanjian, E., Kim, J. & Xu, M.: Environmental, social, and economic implications of global reuse and recycling of personal computers. *Environmental Science & Technology* 42:17 (2008), pp. 6446–6454.

Williamson, O.: Transactions-cost economics: the governance of contractual relations. *Journal of Law and Economics* 22:2 (1979), pp. 233–262.

CHAPTER 6

Increasing green energy market efficiency using micro agreements

Kassidy Clark, Martijn Warnier & Frances Brazier

The recent trend towards increased utilization of green energy sources promises a future of sustainable energy. However, green sources are typically intermittent and unpredictable (e.g. wind power), which leads to increased complexity and, ultimately, decreased efficiency. For example, it is estimated that 90% of the full energy potential of an average wind farm never reaches the end consumer due to poor coordination with consumers (Piwko *et al.*, 2005). To enable better coordination, an energy market using short-term "micro" agreements is required to allow consumers to react in real time to changes in production. This will increase efficient utilization of green sources and reduce energy costs for the consumer. The current market framework cannot adapt quickly to changes due to the constraints of static leases and fixed prices. This means that green energy supply cannot always find green energy demand and vice versa, resulting in low market efficiency. This chapter describes a dynamic market with micro agreements that enables consumers to react to changes in green production.

6.1 INTRODUCTION

Many countries are investing in renewable (or "green") energy sources and new, advanced infrastructure. The European Union has set a goal for its member countries to use renewable sources for at least 20% of total energy consumption by 2020 (European Commission, 2009). In particular, the German government unveiled plans to invest 20 billion euros in a new energy network to support a goal of 80% energy from renewable sources (Luttikhuis, 2012).

As more green energy sources are harnessed, it becomes more difficult for future energy markets to utilize them efficiently. Although there is ongoing research into the use of future electric vehicles for grid energy storage (Peterson *et al.*, 2010) and other techniques, electric energy cannot currently be stored in a cost-effective manner. Therefore, it must be consumed at virtually the same moment it is generated. The balance of energy generation and production is a substantial challenge even when using traditional, continuous energy sources, such as coal power. Green energy sources, such as wind power, further complicate this balance as they are intermittent and unpredictable. For instance, energy demand at the office rarely coincides with weather conditions at the wind park. In fact, it is estimated that most wind farms have effective capacity of 10% of their full potential (Piwko *et al.*, 2005). This means that only 10% of their full potential power reaches consumers, when consumers need it, due to planning and coordination difficulties. Such mismatches between supply and demand result in an inefficient market with substantial wasted potential energy.

One possible solution to the problem of market inefficiency is demand side management (DSM). As the production of energy cannot be shifted in some cases (e.g. wind power), DSM attempts instead to shift demand at the consumer end. Demand shifting occurs when an energy consumer reacts to changes in the energy market (e.g. price, availability) by increasing or decreasing consumption. DSM requires communication, incentive, and action mechanisms. Firstly, the market must communicate with consumers to inform them of the changes in price or (green) availability. Secondly, consumers require an incentive to react to these market signals. One possible incentive mechanism is real-time pricing (RTP) in which prices change in real time to reflect current energy availability.

To facilitate RTP and DSM, this chapter advocates the use of micro agreements. A micro agreement is a short-term agreement directly between parties, in this case, energy providers and energy consumers. A consumer creates an agreement directly with a provider. The agreement specifies the terms and conditions of the services provided. For instance, the price and percentage of green sources for electrical energy. The agreements must be for short periods (e.g. 1 h) so that consumers can quickly react to changes in the market, such as cheaper prices or greener providers, by immediately migrating to new providers. This is opposed to the approach commonly used that uses long-term (e.g. 1 year) contracts with fixed prices, which prevents immediate response to market signals. Micro agreements can increase effective capacity at a wind farm by enabling consumers to immediately react to increased wind energy production by purchasing energy directly from that wind farm.

This chapter describes an energy market framework using micro agreements with an RTP incentive mechanism. Furthermore, this chapter proposes using agent technology to automate negotiation and monitoring of micro agreements to facilitate DSM. The result is a flexible market that quickly reacts to changes in price, availability, and energy type (e.g. green). This reduces wasted overcapacity for the producer, lowers prices for the consumer, and increases overall utilization of green energy sources.

This chapter is organized as follows. Section 6.2 gives an overview of the future energy landscape, including the challenges of intermittent production, the need for DSM and the benefits of RTP. Section 6.3 describes how agent technology can automate market processes, including service negotiation. Section 6.4 explains the role of mico agreements in future energy markets, including the benefits to both consumer and provider. Finally, Section 6.5 concludes the chapter.

6.2 FUTURE ENERGY MARKETS

Global investment in renewable energy has grown by more than 600% since 2004 (McCrone *et al.*, 2011). This trend is fueled in part by the growing concerns for the environmental impact of fossil fuels (e.g. CO_2 emissions). In The Netherlands, the percentage of total energy production from renewable sources has quadrupled in the past decade[1]. In Germany, the government recently unveiled plans to invest 20 billion euros in a new energy network to support a goal of 80% energy from renewable resources (Luttikhuis, 2012). The energy landscape of the future is becoming "greener", but the influx of renewable energy presents a challenge to the current energy grid. These challenges are met with new coordination mechanisms and incentives.

6.2.1 *Intermittent and distributed generation*

Renewable resources typically produce intermittently and are thus more difficult to manage. Sources such as wind and solar power are directly dependent on largely unpredictable weather conditions. Intermittent production complicates power generation and transmission planning, such that a typical wind farm operates with an effective capacity of only 10% of the full capacity (Piwko *et al.*, 2005). This means that only 10% of its full potential power reaches consumers when they need it.

The challenge of intermittent generation is exacerbated by the lack of inexpensive, abundant energy storage technologies. Traditionally, chemical batteries are too small and too expensive to handle the current storage needs. Demand is increasing for wider application of alternative technologies, including mechanical flywheels (Lazarewicz and Rojas, 2004), new battery construction (e.g. liquid metal batteries[2]), and the use of electric vehicles as energy storage (Peterson *et al.*, 2010). However, these technologies have not yet reached wide-spread adoption.

---

[1]Centraal Bureau voor de Statistiek. http://www.cbs.nl/nl-NL/menu/themas/industrie-energie/publicaties/artikelen/archief/2010/2010-3105-wm.htm. Accessed: September 2012.

[2]Clean Technica. http://cleantechnica.com/2012/02/20/mit-liquid-batteries-for-utilities-could-make-renewables-competitive-and-it-is-not-lithium-ion/. Accessed: June 2012.

Distribution of energy generation will also change with the influx of green technologies. Rather than the current paradigm of centralized power stations, a larger percentage of power will be generated by distributed resources. For instance, consumers install solar panels and micro combined heat and power (micro CHP)[3]. Currently, any overcapacity generated by these distributed resources is sold back to a single retailer. This retailer is then confronted with the complex task of reselling and redistributing this sporadically generated energy. One method to reduce the complexity on centralized management (e.g. retailer) is to form groups of small-scale producers and consumers. These groups are also referred to as microgrids or virtual power plants (VPPs). Members of these groups supply each others' energy demands, independently from the rest of the energy grid (Hatziargyriou *et al.*, 2007; Ramchurn *et al.*, 2012).

To cope with the added complexity of intermittent energy sources and the distribution of production, a new energy management approach has been proposed: the Smart Grid (Gellings, 2009; Ramchurn *et al.*, 2012). Two key attributes of the Smart Grid are (1) bi-directional flow of information and (2) DSM. Information flow includes real-time metering of user consumption (e.g. smart meters[4]) such that producers can quickly see and respond to changes in demand. Information flow also includes providing consumers with real-time data about energy availability (e.g. is green energy being produced at the local wind farm right now) and price (e.g. demand is low, thus the current price is reduced). Information flow is critical to enabling consumer participation.

6.2.2 *Demand side management*

DSM is a technique of dealing with fluctuations in energy production by modifying demand to match supply (Gottwalt *et al.*, 2011; Strbac, 2008). Traditionally, the complex task of matching supply to demand is handled entirely by the producers of energy. DSM enables end consumers to participate in this task by temporally shifting energy consumption in response to signals from the market, such as price. Typically, economical incentives are used to encourage consumers to reduce consumption during peak periods, when energy is scarce and expensive. One form of DSM is *time-of-use schemes* that offer different energy prices based on time of day. Typically, nighttime prices are substantially lower than daytime prices due to the demand of businesses during working hours. A consumer is thus economically rewarded for running washing- and drying machines during the night rather than during the day. Unfortunately, time-of-use schemes have unwanted side effects that result in significant usage peaks as soon as the lower tariff period begins (Ramchurn *et al.*, 2012). A more dynamic approach is required.

DSM offers many benefits, including the reduction of overcapacity. Currently, energy producers generate approximately 20% more energy than is required (Strbac, 2008). This overcapacity is needed to handle unforeseen peaks in demand or problems with production. DSM encourages consumers to avoid creating such peaks and thus overcapacity can be reduced. This has been shown to reduce wasted energy and costs (Ramchurn *et al.*, 2012).

Another benefit of DSM is the increased utilization of intermittent sources. Renewable sources of energy, such as wind power, produce power intermittently, as opposed to the consistent power provided by traditional sources, such as coal. Intermittent production means that energy is not produced in a controllable way that matches consumption patterns, but rather is produced sporadically (e.g. depending on the weather and so on). Due to the unpredictable production schedule of intermittent sources, very little of the power generated actually reaches the end consumer when it is needed. As mentioned above, wind farms have an estimated effective capacity of only 10% (Piwko *et al.*, 2005). This means that only 10% of the full energy potential of wind power actually reaches the end consumer. The rest of the energy is lost due to lack of demand at the moment of production.

---

[3] Micro CHP is a small gas powered turbine installed in a residential area, near the end customers, that converts gas into both electricity and heat. Performing the conversion near the end users reduces transmission losses and allows efficient use of heat produced as a by-product of conversion (De Paepe *et al.*, 2006).

[4] A smart meter is a device to measure electricity usage and communicate usage information to the consumer and/or energy provider in (semi) real-time (Venables, 2007).

6.2.3   *Real-time pricing*

Currently, consumers sign long-term contracts with energy providers for a fixed price per kilowatt hour (kWh). This approach stabilizes prices for consumers, but lacks the incentives necessary for consumer initiated demand shifting. An alternative to this approach is RTP. Under this scheme, consumers pay the current price of energy as determined by conventional market forces (e.g. supply and demand). When demand is high, the price of energy increases and vice versa. In this way, RTP offers a tangible incentive to reduce consumption during peak demand periods and shift this consumption to low demand periods. The intended results of RTP are (1) increase market efficiency (e.g. matching supply and demand) (2) reduce overall energy production, and (3) empower consumers to take an active roll in reducing energy costs.

Research shows (Borenstein, 2002; Cramton, 2003) that RTP can be effective at increasing market efficiency. In fact, simulations have shown that RTP is more than five times as efficient as time-of-use schemes (Borenstein, 2005). If consumers are aware of the current price of electricity (e.g. hourly), they tend to reduce consumption when electricity is expensive and shift their usage to periods when prices are cheap. If these prices reflect the current market conditions, such as current production capacity and consumer demand, then the result is that consumers reduce demand when supply is limited and increase when supply is abundant. Matching supply to demand reduces wasted energy (e.g. produced but unconsumed) and increases overall market efficiency.

Several energy providers have offered RTP to their customers on a voluntary basis with varying results (Barbose *et al.*, 2004). In general, the goals of reducing price and overall production load were met. On average, RTP schemes reduced energy load between 12 and 33%. However, most customers failed to respond to hourly changes in price, unless the price rose above a certain threshold. In most cases, the shortcomings of the RTP scheme can be traced to implementation faults rather than a fundamental problem with the theory. For instance, many consumers lacked user-friendly feedback mechanisms (e.g. smart meter) or felt that the correlation between prices and usage was not transparent and therefore lacked trust in the scheme.

An often cited failure of RTP was the California energy crisis of 2000. This application of RTP led to extreme price volatility with disastrous consequences. Researchers have since concluded that situation was the result of a poor market design that allowed producers to exercise market power by artificially reducing supply to increase prices (Borenstein, 2002).

When applied correctly, RTP has been shown to (1) stabilize demand and simplify the task of production management and (2) increase market efficiency and reduce waste (e.g. supply overcapacity) (Borenstein, 2005). In fact, RTP was shown to be five times as efficient as time-of-use schemes (e.g. day- and night-time tariffs). Price volatility can be mitigated using a combination of long-term contracts between wholesalers and RTP for end users (Borenstein, 2002) or a two-part rate scheme that combines an energy quota at a fixed price with RTP for deviations from that quota (Barbose *et al.*, 2004).

Automation technologies that can coordinate information flows and assist consumers with DSM currently exist. Modern smart meters combined with intelligent agent technology can automate the tasks of (1) monitoring energy prices, (2) negotiating (e.g. double auctions), and (3) coordinating demand shifting (Deindl *et al.*, 2008). This will alleviate the burden of manually reacting to RTP to efficiently reduce costs and better utilize green energy. These automation techniques are discussed below.

6.3   MARKET AUTOMATION

Online markets exist for many areas of commerce, including web (Cheng *et al.*, 2007), grid (Buyya and Vazhkudai, 2001), cloud services (Buyya *et al.*, 2008), and (industrial) Energy Auctions. An online marketplace is essentially a location where providers can electronically advertise their services and consumers can access those services. A marketplace can offer additional services and structure, such as an explicit ontology, terminology, protocols, and facilitation. Facilitation may include assistance finding services, negotiating a price, or resolving a conflict.

Online markets are organized such that consumers can compare services and choose between competing providers. A standardized language is used to define (compositional) services. Standardized protocols are used to make it straightforward to switch seamlessly between providers. A marketplace typically offers some type of directory service, where providers can publish available services. Third-party brokers can be used to actively match customers to appropriate providers. Agreements reached between parties can be formalized using service level agreements (SLA) as described in Section 6.4.

Many market processes can be (partially) automated with technologies, such as software agents. These techniques can be applied to monitor and respond to real-time information quickly and efficiently. Removing the need for constant human interaction makes it possible to increase the speed and frequency of market interactions. In addition, intelligent automation can often react faster than humans in complex, dynamic systems that can be difficult for humans to understand and follow. Another crucial market process is negotiation. After a consumer discovers a provider with a particular service (e.g. wind energy), the two parties much reach an agreement regarding the terms and conditions of the service, including price and quality of service (QoS) (e.g. uptime, time to repair an outage, and minimum green percentage). In recent years, much research has been done on the area of applying agent technology to automate the process of negotiation.

6.3.1 *Automated agent-based negotiation*

Software agents are one approach to modeling users in online markets. In this chapter, a software agent is defined as: a (semi) autonomous, mobile piece of software capable of independent action and communication. A user can program a software agent to act on his/her behalf to make decisions based on his/her preferences. This technology is particularly well-suited to the task of automating the service negotiation process (Chao *et al.*, 2002; Debenham and Lawrence, 2006; Jennings *et al.*, 2001).

Agents have been proposed to model and manage complex, distributed systems, such as the energy market (Duan and Deconinck, 2010; Koritarov, 2004; Vytelingum *et al.*, 2011). Separate agents represent each unique role of the energy market, including energy consumer, producer, mediator, and broker. In some cases, agents take on dual roles. For instance, a residential consumer becomes a producer if solar panels are installed on his/her roof. Additional agents can be used to represent transmission companies, distribution companies, and independent system operators.

Agents in these systems consist of personal preferences, internal decision rules, and some means to store and process historical actions. Personal preferences can include things such as *favor local producers* or priority lists such as *solar, wind, biomass, nuclear*. Internal decision rules allow an agent to access the situation (e.g. read an offer) and take action (e.g. reject the offer or propose a counter-offer). An important instance of decision rules is *negotiation strategy*. This is the strategy that guides the negotiation process and determines when to accept, reject, or counter an offer. Finally, an agent typically has some method to store and process past actions. For instance, consistently poor negotiation performance may cause the agent to modify the negotiation strategy.

An important prerequisite for user acceptance of an agent-based system, especially when used for critical tasks, is that users trust the system (Hsu, 2008; Huang *et al.*, 2008; Manchala, 1998). Transparency enhancements, auditing mechanisms, and third-party certification can be built into the system to promote trust and acceptance.

6.3.2 *Automated energy market*

In future energy markets, agents represent consumers and providers. A residential consumer is represented by an agent that has access to all internal consumer data, including historical energy usage patterns. In addition, this agent has (limited) control of energy consumption in the home, for instance, access to "smart" appliances such as refrigerators or clothes dryers. At a minimum, the agent must be able to monitor energy usage, turn on, and turn off the appliance via some type of network. Additional intelligence can indicate the priorities and special requirements of individual appliances. For instance, a refrigerator can safely be turned off for a short period of time without any serious consequences. However, it may be unacceptable if the television were to turn off in the middle of a show.

The consumer agent finds the best deal among energy providers, given a set of preferences. An example of preferences could be (1) minimize price and (2) maximize green energy. The consumer agent surveys the marketplace to find providers offering suitable services. If one or more suitable providers is found, the agent negotiates the terms and conditions of service. If an agreement is reached an SLA is created (see Section 6.4). The provider then begins service provisioning. During the lifetime of the agreement, the agent monitors the service to ensure that the terms and conditions are met. The monitoring process builds a secure audit log of all transactions. If a violation is detected, penalties are applied. In the case of a dispute, the audit log is consulted to resolve the conflict and advise appropriate action. The entire process is repeated regularly (e.g. every hour) to ensure a consumer has the best price and service.

The consumer agent acts as a smart energy gateway for each end user (e.g. home or office building). This agent interacts directly with agents that represent producers of energy. For instance, one agent represents a company controlling a coal power plant and another agent represents a wind farm. These agents have access to all internal producer data, including current production capacity, profit margins, and current demand. The producer agents negotiate sales to consumers based on this local data and their own negotiation strategy (e.g. maximize usage, maximize profit).

Energy market automation is essential to respond effectively to changes in the market, such as lower prices or abundant green energy, as these changes occur often and cannot be predicted. Automation supports effective DSM in real-time. For instance, based on market conditions, an agent may choose to postpone energy consumption (e.g. clothes dryer). In this way, green energy production can better find green energy demand.

6.4   MICRO AGREEMENTS

Future energy markets use Micro Service Level Agreements (micro-SLAs) to facilitate frequent renegotiation of energy agreements. A micro-SLA is a short-term agreement between a consumer and a provider that specifies the services to be provided and the applicable service quality guarantees (QoS). In order to react to RTP, a consumer creates a micro-SLA with a an energy provider for a short period (e.g. 1 h). When this micro-SLA expires, the consumer reevaluates its energy needs, surveys the energy prices from competing providers, and negotiates a new micro-SLA. If green energy is available (e.g. the local wind park is at full capacity), the consumer negotiates a micro-SLA with the provider of that energy. If the consumer finds all current energy offers too expensive, the consumer can consider shifting its energy demand.

Micro agreements create a flexible environment where consumers can actively favor green energy resources over other resources. Consumers can utilize as much green power as possible by maximizing demand during green peak periods. This can have a secondary effect of financially rewarding additional investment by producers in green energy resources.

6.4.1   *Service level agreements*

SLAs are agreements between multiple parties to specify terms of service. They involve at least one service provider and at least one service consumer and specify the services to be provided. Additionally, these documents can specify QoS guarantees; payments for compliance, or penalties for violation of the agreement. QoS attributes can be expressed as a set of (name, value) pairs where name refers to a service level objective (SLO) and value represents the requested level of service. An SLO specifies the particular object to be measured, how measurements are performed, and any actions that take place after measurement.

To illustrate an SLA, consider the following example. A consumer agrees to purchase Internet connectivity from an Internet service provider. They two parties create an SLA that contains

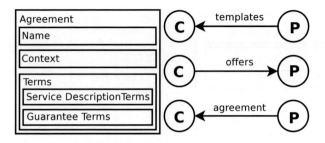

Figure 6.1. WS-Agreement SLA specification and protocol.

the following information. First, the SLA specifies the service as *electricity*. Then, the SLA describes the QoS guarantees that apply to this service, in the form of SLOs. These may include *(greenness, >= 50%)* or *(uptime, >= 99%)*. The SLA also explains how each of these SLOs are measured by testing at hourly intervals. Finally, the SLA states that the provider is penalized financially for each violation.

Traditionally, SLAs are written and signed between legal entities, representing each of the parties involved. In recent years, much attention has been given to automating this negotiation process (Ludwig *et al.*, 2006; Padget *et al.*, 2005; Sahai *et al.*, 2002). Several specifications have been proposed for this purpose, including WS-Agreement (WSAG) (Andrieux *et al.*, 2011).

WSAG defines a language with which to express agreements and a base protocol to advertise and negotiate services, as depicted in Figure 6.1. Agreements are expressed in structured XML documents. These documents contain the name, context (e.g. the parties involved), and the terms. Each term contains a service description (e.g. greenness) and an optional guarantee (e.g. >= 50%).

Initially, when a consumer *(agreement initiator)* requests an indication of services available from a provider *(agreement responder)*, the provider produces an overview of available services *(templates)*. Based on this information, the consumer proposes an SLA *(offer)*. The provider decides whether or not to accept the proposed SLA. If accepted, service provisioning begins *(lease)*.

The WS-Agreement-Negotiation (WSAN) specification (Waldrich *et al.*, 2011) extends the WSAG standard. This extension separates the process of agreement creation using WSAG from the process of negotiating the terms of agreements using WSAN. The key extensions of WSAN are (1) renegotiation of existing agreements, (2) multi-round negotiation sessions, (3) parallel negotiation sessions, and (4) asymmetry of initiation and response actions.

Negotiation sessions use offers expressed in XML and follow a series of offers and counter offers, as depicted in Figure 6.2. In addition to the components used in WSAG, WSAN documents include a *negotiation offer ID* and *negotiation constraints*. These constraints define restrictions for the structure and values used in all offers of a negotiation session. For instance, a provider may specify that price is nonnegotiable or is limited to values greater than 1000.

Negotiation sessions are a series of offers and counter offers. Multiple sessions may be executed in parallel. Beginning with a *Template* a party creates an offer, following the WSAG protocol. The counter party may respond to this offer by accepting, rejecting, or proposing a counter offer. Offers and counter offers are encapsulated in negotiation messages. The protocol specifies that negotiation messages contain metadata, including *offer ID, counter offer to ID,* and *creator ID*. This metadata allows the messages to be organized into a tree, as depicted in Figure 6.2.

A counter offer is typically in response to the offer that immediately precedes it. However, a counter offer may also be in response to an earlier offer. For instance, if a party decides that a branch of the tree is no longer worth pursuing, that party creates a new counter offer to an earlier offer, thus creating a new branch of the tree. This is indicated as a "rollback" in Figure 6.2.

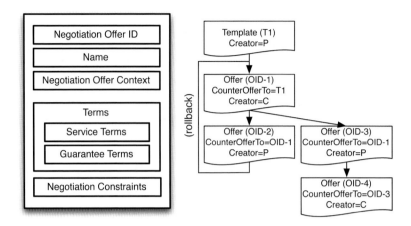

Figure 6.2.   WS-agreement-negotiation offer specification and messages.

6.4.2   *Benefits to the consumer*

In this context, a consumer is defined as the end user of electricity in the traditional market configuration. This includes residential homes and industrial buildings.

Elimination of static, long-term agreements with a single provider allows consumers to freely choose the best offer for energy. This creates true market competition between providers offering the lowest price or best quality product (e.g. highest renewable percentage). Micro-SLAs increase the buying power of a single consumer by obliging providers to respond to consumer preferences. For instance, a small, independent energy provider produces 100% green energy. Consumers can choose to immediately migrate to this provider. This sends a signal that consumers prefer this energy option. Other providers are obliged to offer similar products to entice consumers to return.

Micro-SLA also increases the selling power of consumers with local energy production (e.g. solar cells on a residential roof). Instead of the current, obligation to sell overcapacity back to a single provider for a fixed price, consumers can offer their energy to the marketplace at large. Consumers may sell their overcapacity to a selection of providers or even to other (local) consumers. In this context, the consumer agent assumes the role of salesman and attempts to find the highest bidder. The possibility of reselling energy may further motivate consumers to reduce consumption during peak periods, because energy sales during these periods will fetch an even higher price.

6.4.3   *Benefits to the producer*

In this context, a producer is defined as the generator of electricity in the traditional market configuration. This includes, centralized power plants such as coal, gas, and nuclear plants, as well as off-shore wind farms and concentrated solar farms.

One benefit to the producer is the reduction of overcapacity. When consumers are able to react to price fluctuations, consumption during peak periods is reduced and less overcapacity is required (Borenstein, 2005; Ramchurn *et al.*, 2012). This reduces wasted energy resources and increases profit margins.

An additional long-term benefit that stems from the reduction of overcapacity is a decrease in capital expenditure (CAPEX) and operating expense (OPEX). Consistent, long-term reduction of overcapacity requirements equates to reduced investment in production capability. For instance, an additional turbine used to produce 20% of total capacity will not need to be built and an existing one can be decommissioned earlier than expected.

Smaller producers with 100% green, intermittent sources can compete more easily in a market place with short-term contracts. In the traditional market, an energy producer with intermittent sources (e.g. wind) also requires a portfolio of stable energy sources (e.g. coal) to guarantee its consumers continuous power. This requires significant investment and thus smaller producers sell to other retailers rather than directly to the end consumer. With micro-agreements the owner of a wind farm can sell directly to customers dependent on how much energy is being produced due to weather conditions. When production is high, the client base is increased dynamically and vice versa. This increases revenue and reduces wasted potential energy. Furthermore, allowing consumers to react to changes in energy prices gives more market power to small, specialized producers (Cramton, 2003).

Another benefit to producers is direct access to user preferences. With micro-agreements, consumers can "vote" with their pocketbooks for their energy preferences in real-time. Producers can clearly follow consumer trends and react by investing accordingly. Typically such data are collected through limited surveys or analyzing trends over longer periods, such as years or decades.

6.5 CONCLUSION

In recent years, much has been invested in adding green energy sources to the energy grid. Unfortunately, due to planning difficulties, these resources are significantly underutilized. Green energy supply does not efficiently reach consumers when they need it. Future energy markets require several changes to increase this efficiency using DSM. First, micro agreements must replace long-term contracts. Micro agreements allow consumers to quickly react to changes in the market, such as price and availability. Second, RTP must be used to give consumers incentive to react to market signals by shifting demand. For instance, reducing consumption when wind power generation is scarce and increasing consumption when it is abundant.

Studies have shown the effectiveness of RTP at encouraging DSM in energy markets. Current automation techniques, including software agents, can further increase this effectiveness by reducing the manual burden on the consumer. Future energy markets could include automated agents that represent each consumer and provider. These agents interact, negotiate and create micro agreements with one another. This new market structure reduces wasted energy (e.g. overcapacity), lowers prices, and increases utilization of green resources.

Future work will simulate this marketplace using software agents, the WSAN negotiation protocol, and real-world usage statistics.

ACKNOWLEDGEMENTS

This work is supported by the NLnet Foundation (www.nlnet.nl).

REFERENCES

Andrieux, A., Czajkowski, K., Dan, A., Keahey, K., Ludwig, H., Nakata, T., Pruyne, J., Rofrano, J., Tuecke, S. & Xu, M.: Web Services Agreement Specification (WS-Agreement). *Global Grid Forum GRAAP-WG*, 2011.

Barbose, G., Goldman, C. & Neenan, B.: A survey of utility experience with real time pricing. Technical report, Ernest Orlando Lawrence Berkeley National Laboratory, Berkeley, CA, 2004.

Borenstein, S.: The trouble with electricity markets: understanding california's restructuring disaster. *The Journal of Economic Perspectives* 16:1 (2002), pp. 191–211.

Borenstein, S.: The long-run efficiency of real-time electricity pricing. *Energy Journal* 26:3 (2005), pp. 93–116.

Buyya, R. & Vazhkudai, S.: Compute power market: towards a market-oriented grid, ccgrid. Published by the *IEEE Computer Society*, 2001, pp. 574–581.

Buyya, R., Yeo, C. & Venugopal, S.: Market-oriented cloud computing: vision, hype, and reality for delivering it services as computing utilities. *Proceedings of the 10th IEEE International Conference on High Performance Computing and Communications, 2008, HPCC'08,* IEEE, Dailan, China, 2008, pp. 5–13.

Chao, K.-M., Anane, R., Chen, J.-H. & Gatward, R.: Negotiating agents in a market-oriented grid. *Proceedings of the 2nd IEEE/ACM International Symposium on Cluster Computing and the Grid,* Berlin, Germany, 2002, p. 436.

Cheng, S., Chang, C., Zhang, L. & Kim, T.: Towards competitive web service market. *Proceedings of the 11th IEEE International Workshop on Future Trends of Distributed Computing Systems, 2007, (FTDCS'07),* IEEE, Sedona, Arizona, USA, 2007, pp. 213–219.

Cramton, P.: Electricity market design: the good, the bad, and the ugly. *Proceedings of the 36th Annual Hawaii International Conference on System Sciences,* 2003, IEEE, Hawaii, USA, 2003, pp. 8–16.

De Paepe, M., D'Herdt, P. & Mertens, D.: Micro-chp systems for residential applications. *Energy Conversion and Management* 47:18 (2006), pp. 3435–3446.

Debenham, J. & Lawrence, E.: Negotiating agents for Internet Commerce. *Proceedings of the International Conference on Computational Intelligence for Modelling, Control and Automation, 2006 and International Conference on Intelligent Agents, Web Technologies and Internet Commerce,* Sydney, Australia, 2006, p. 269.

Deindl, M., Block, C., Vahidov, R. & Neumann, D.: Load shifting agents for automated demand side management in micro energy grids. *Proceedings of the Second IEEE International Conference on Self-Adaptive and Self-Organizing Systems (SASO'08),* Venice, Italy, 2008, pp. 487–488.

Duan, R. & Deconinck, G.: Future electricity market interoperability of a multi-agent model of the smart grid. *Proceedings of the IEEE International Conference on Networking, Sensing and Control (ICNSC'10),* IEEE, Chicago, Illinois, USA, 2010, pp. 625–630.

European Commission. Investing in the development of low carbon technologies (set-plan). *COM (2009)* 519, 2009.

Gellings, C.: *The smart grid: enabling energy efficiency and demand response.* CRC Press, Boca Raton, FL, 2009.

Gottwalt, S., Ketter, W., Block, C., Collins, J. & Weinhardt, C.: Demand side management—a simulation of household behavior under variable prices. *Energy Policy* 39:12 (2011), pp. 8163–8174.

Hatziargyriou, N., Asano, H., Iravani, R. & Marnay, C.: Microgrids. *Power and Energy Magazine, IEEE* 5:4 (2007), pp. 78 –94.

Hsu, C.: Dominant factors for online trust. *Proceedings of the International Conference on Cyberworlds,* Hangzhou, China, 2008, pp. 165–172.

Huang, H., Zhu, G. & Jin, S.: Revisiting trust and reputation in multi-agent systems. *Proceedings of the International Colloquium on Computing, Communication, Control, and Management, 2008. CCCM'08. ISECS,* Vol. 1, Guangzhou, China, 2008.

Jennings, N., Faratin, P., Lomuscio, A., Parsons, S., Wooldridge, M. & Sierra, C.: Automated negotiation: prospects, methods and challenges. *Group Decision and Negotiation* 10:2 (2001), pp. 199–215.

Koritarov, V.: Real-world market representation with agents. *Power and Energy Magazine, IEEE* 2:4 (2004), pp. 39–46.

Lazarewicz, M. & Rojas, A.: Grid frequency regulation by recycling electrical energy in flywheels. *Power Engineering Society General Meeting 2004,* IEEE, Vol. 2, Denver, Colorado, USA, 2004, pp. 2038–2042.

Ludwig, A., Braun, P., Kowalczyk, R. & Franczyk, B.: A framework for automated negotiation of service level agreements in services grids. *Lecture Notes in Computer Science* 3812 (2006), pp. 89–101.

Manchala, D.: Trust metrics, models and protocols for electronic commerce transactions. *Proceedings of the 18th International Conference on Distributed Computing Systems,* Amsterdam, The Netherlands, 1998, pp. 312–321.

McCrone, A., Usher, E., Sonntag-O'Brien, V., Moslener, U., Andreas, J. & Gruening, C.: Global trends in renewable energy investment 2011. Technical report, United Nations Environment Programme (UNEP) and Bloomberg New Energy Finance, New York, New York, USA, 2011.

Padget, J., Djemame, K. & Dew, P.: Grid-based SLA management. *Lecture Notes in Computer Science* 3470, 2005, pp. 1076–1085.

Peterson, S.B., Whitacre, J. & Apt, J.: The economics of using plug-in hybrid electric vehicle battery packs for grid storage. *Journal of Power Sources* 195:8 (2010), pp. 2377–2384.

Piwko, R., Osborn, D., Gramlich, R., Jordan, G., Hawkins, D. & Porter, K.: Wind energy delivery issues [transmission planning and competitive electricity market operation]. *Power and Energy Magazine, IEEE* 3:6 (2005), pp. 47–56.

Ramchurn, S., Vytelingum, P., Rogers, A. & Jennings, N.: Putting the "smarts" into the smart grid: a grand challenge for artificial intelligence. *Communications of the ACM* 55:4 (2012), pp. 86–97.

Sahai, A., Machiraju, V., Sayal, M., van Moorsel, A., Casati, F. & Jin, L.: Automated SLA monitoring for web services. *Lecture Notes in Computer Science* Volume 2506 (2002), Springer Verlag, Berlin, Germany, pp. 28–41.

Strbac, G.: Demand side management: benefits and challenges. *Energy Policy* 36:12 (2008), pp. 4419–4426.

Venables, M.: Smart meters make smart consumers. *Engineering Technology* 2:4 (2007), p. 23.

Vytelingum, P., Ramchurn, S., Voice, T., Rogers, A. & Jennings, N.: Agent-based modeling of smart-grid market operations. *Power and Energy Society General Meeting, 2011 IEEE*, IEEE, 2011, Detroit, Michigan, USA, pp. 1–8.

Waldrich, O., Battre, D., Brazier, F.M.T., Clark, K.P., Oey, M.A., Papaspyrou, A., Wieder, P. & Ziegler, W.: WS-Agreement Negotiation: Version 1.0 (GFD-R-P.193). Technical report, Open Grid Forum, Grid Resource Allocation Agreement Protocol (GRAAP) WG, 2011.

CHAPTER 7

Framework for measuring environmental efficiency of IT and setting strategies for green IT: A case study providing guidance to chief information officers

Johanne Punte Kalsheim & Erik Beulen

7.1 INTRODUCTION

Sustainability has become one of the greatest challenges of society today. To companies sustainability is essential to the long-term successfulness of business operations and is often driven by factors such as costs, new market opportunities, ethical responsibility and imago (Dao, *et al.*, 2011; Porter and Kramer, 2006; Schaltegger, 2008; Unhelkar, 2011). A result of this has been increasing attention to sustainability in several industries.

IT decision-makers want to contribute to environmental sustainability and the corporate responsibility strategy of an organization, but how to do this is often unclear. Corporate responsibility (CR) strategy is defined as sustainable business activities as well as corporate social investments and programs. The relationship between CR strategy and green IT will be further elaborated on in the next section. Important to notice is that organizations that emphasize sustainability may not always extent their environmental efforts to the IT department (Cone, 2006). A framework providing insights into the greenness (relative sustainability) of the hardware IT infrastructure of an organization is an important step towards creating awareness and supports IT decision-makers in their contribution to the overall CR strategy of an organization. Consequently, the aim of the research was to design a framework that could assess the relative greenness of the hardware IT infrastructure of an organization. Relative greenness indicates that achieving sustainability is a process. A framework is then needed to measure progress and help select the most effective measures. The hardware IT infrastructure is defined as the shared physical IT equipment within an organization that is used to process, store or transmit computer programs or data (Linberg, 1999). The focus of the research was on this part of the total ICT-infrastructure that also consists of software. The research question addressed in this chapter is therefore formulated as follows: *What generic framework could be developed to assess the relative greenness of the hardware IT infrastructure of an organization as a step towards a comprehensive corporate responsibility (CR) strategy?*

To address the research question we define green IT in the following section where we also link Green IT to the concept of Corporate Responsibility followed by an elaboration on performance indicators for green IT. This is followed by a set of design requirements for a green IT framework that we tested in a case study. This chapter then proceeds with an evaluation and reflection on the outcomes followed by recommendations on implementation of the green IT framework and ends with conclusions and further suggestions for further research.

7.2 GREEN IT

This section details the definition of *green IT* as well as the relationship between *green IT* and corporate responsibility strategy.

7.2.1   *Green IT definitions*

Green IT has become a catchphrase in IT management even though a common understanding of the scope and coverage of the concept is missing in research and practice (Velte *et al.*, 2008). Some researchers also prefer the term *green Information Systems (IS)* as this concept incorporates *green IT* and comprise a greater variety of possible initiatives to support sustainable business processes beyond IT (Watson *et al.*, 2010). A comprehensive overview of *green IT* definitions is provided by Molla (2009). From the various concepts the author suggests *green IT* is:

> "*a systematic application of ecological-sustainability criteria (such as pollution prevention, product stewardship, use of clean technologies) to the creation, sourcing, use and disposal of the IT technical infrastructure as well as within the IT human and managerial components of the IT infrastructure, in order to reduce IT, business process and supply-chain related emissions, waste and water use; improve energy efficiency and generate Green economic rent*" (Molla, 2009:757)

Molla's definition indicates that *green IT* requires a holistic approach that incorporates the entire life cycle of the IT infrastructure (Molla, 2009). That the life cycle of IT is a significant part of *green IT*, is emphasized by several other authors (Capra and Merlo, 2009; Dao *et al.*, 2011; Elliot, 2011; Harmon and Auseklis, 2009; Harmon and Demirkan, 2011; Harmon, *et al.*, 2010; Hird, 2010; Li and Zhou, 2011; Molla and Cooper, 2009; Murugesan, 2008; Unhelkar, 2011). However, where the life cycle of IT starts and ends is either unclear or differs in scope in literature on *green IT*. Murugesan suggest the life cycle of IT entails design, manufacture, use and disposal, whereas Molla and Coopers (2009) incorporates sourcing, operation, recycling, reuse and disposal.

According to the study by Molla (2009) there are several similar terms in *green IT* definitions such as *green, sustainability, environmental sustainability, eco-efficiency* (i.e. costs and resource efficiency)*, life cycle, human and managerial components* and *greening by IT*. *Of importance* is a distinction between greening IT and greening by IT. The focus of greening IT is on greening operations, whereas the greening by IT focus is on using IT services to green an entire organization (Hird, 2010). Both ways could be seen a first step towards the goal of sustainability (Molla, 2009). Various concepts related to *IT, green* and *green IT* have been defined to understand what the concept of *green IT* entails. These are summarized in Table 7.1.

We offer the following synthesis: *management of the environmental sustainability and eco-efficiency over the life cycle of IT (greening IT) and the use of IT services to meet overall environmental sustainability goals of an organization* (greening by IT) (Capra and Merlo, 2009; Dao *et al.*, 2011; Elliot, 2011; Harmon and Auseklis, 2009; Harmon and Demirkan, 2011; Harmon, *et al.*, 2010; Hird, 2010; Lamb, 2009; Li and Zhou, 2011; Molla, 2009; Molla and Cooper, 2009; Murugesan, 2008; Unhelkar, 2011). This definition of *green IT* emphasizes the environmental and economic pillars of sustainability. In the next section we elaborate further on green IT by addressing the relationship between the concept and CR strategy of organizations.

7.2.2   *Green IT as a step towards corporate responsibility*

To understand how green IT and Corporate Responsibility (CR) strategy are affiliated, the concept of CR strategy should be addressed first. What is CR strategy? Essentially the ultimate goal of CR is to balance the Triple Bottom Line (3BL) interest beyond an organization's economic interests and that required by law (Linnanen and Panapanaan, 2002; Lo and Sheu, 2007; McWilliams and Siegel, 2001; Wempe and Kaptein, 2002) CR strategy is furthermore about choosing a unique position with respect to this balance to unlock shared value and strengthen company competitiveness (Kramer and Porter, 2006; McWilliams and Siegel, 2001). This balance may be found in corporate sustainability reports (Gallego, 2006; Roca and Searcy, 2012; Skouloudis *et al.*, 2010).

With regards to *green IT* and CR strategy a second "wave" has emerged referred to as *sustainable IT*. In essence, the main driver of this wave is CR (Harmon *et al.*, 2010). Hence, *sustainable*

Table 7.1.   Summary of key green IT concepts.

Concept	Definition from literature	Relevance for Green IT framework
IT	Computing technology for processing and storing ICT for transmitting information (Martin *et al.*, 2009)	Provides a conceptual foundation of the width of the IT and IT infrastructure within an organization next to the hardware IT infrastructure scope chosen.
IT infrastructure	The technical IT-, managerial capability- and IT human infrastructure of an organization (Molla *et al.*, 2011)	
Technical IT infrastructure	The tangible IT resources such as communications technologies, shared services and business applications that utilize the shared infrastructure and form the backbone for business applications (Broadbent and Weil, 1997; Da Silva and e Abreu, 2010; Duncan, 1995)	Provide a conceptual foundation of the various IT resources that could be found within an organization. Expanding the framework, IT software and services should be included.
Hardware IT infrastructure	The shared physical IT equipment within an organization that is used to process, store or transmit computer programs or data (Linberg, 1999).	Hardware IT infrastructure show there are several physical IT equipments in an organization that should be taken into account when developing the Green IT framework
Sustainability	Development that meets the needs of the present world, without compromising the ability of future generations to meet their own needs (Brundtland, 1987)	Sustainability provides the conceptual foundation on the role of IT in sustainable development and position the Green IT framework in a larger sustainability contex
Life cycle thinking	The major activities in the course of a product's life-span from raw material acquisition, manufacture, use and maintenance to the final disposal (SAIC, 2006)	The life cycle approach implies that an organization needs to view Green IT in terms of the entire IT lifecycle. The scope is limited to procurement, use and disposal
Eco-efficiency	Value creation or reduction of costs for the environmental improvement investigation (Huppes and Ishikawa, 2005)	Improving the measurability of green IT to support efficiency of setting priorities of measures.

IT would require implementing a balanced economic, environmental and social responsibility (3BL) strategy beyond an organization's economic interests and that required by law (Linnanen and Panapanaan, 2002; Lo and Sheu, 2007; Wempe and Kaptein, 2002). However, *green IT* as a separate concept from *sustainable IT,* would not be able to achieve this balance due to the lack of emphasis on social aspects of sustainability. On the other hand, strategies aimed at *green IT* could create a unique position in the balance of the 3BL. According to Porter (1996) strategy is always a question of choice, coming along with tradeoffs. This is also the case for CR strategy.

Orsato (2009) suggest there are four types of generic, choice-based competitive environmental strategies that might be relevant when pursuing to green the hardware IT infrastructure of an organization. These will be elaborated on with regards to CR strategy defined by Labuschagne *et al.* (2004) and McWilliams and Siegel (2001).

Eco-efficiency strategies aim at minimizing waste, by-products and emissions. This way, production efficiency is enhanced and costs are reduced (Erek *et al.*, 2011). The definition of *green IT* in the previous section suggests eco-efficiency is a central part of its scope. However, eco-efficiency does not lead to a competitive advantage via differentiation and does not go beyond immediate economic interests of an organization This is the same for *environmental cost leadership*, a strategy striving for radical product innovations through for instance substitution of input materials and business practices (Erek *et al.*, 2011). This may however contribute to the CR strategy when an organization substitutes materials that are hazardous or have a big ecological footprint with more sustainable materials beyond requirements defined in laws and regulations. The strategy may subsequently be entitled *beyond compliance leadership*. However, *beyond compliance leadership* focus on differentiation by pursuing unprofitable environmental initiatives (Erek *et al.*, 2011). This is also a strategy that may contribute to the overall CR of an organization because it strives to go beyond already established laws and regulations. The last strategy, *eco-branding* strives for competitive differentiation through environmental product characteristics that a customer might be willing to pay for (Erek *et al.*, 2011). For suppliers of hardware IT this might create competitive advantage, however also forms an opportunity of lowering the environmental footprint of the organization when measuring this in line with the ISO 14031 standard for environmental performance. *Eco-branding* may contribute to the CR strategy of an organization when it goes beyond and above laws and regulations.

Green IT is also driven by a wish to pursue CR within organizations (Molla, 2009a; 2009b). Harmon and Demirka (2011) state some organization are taking *green IT* to the next level when striving to balance the 3BL, and thereby also assuring a more comprehensive contribution to the CR strategy. IBM has for instance approached *sustainable IT* from an enterprise wide corporate sustainability perspective with initiatives addressing energy conservation, increases in the use of renewable energy and increases in supply chain efficiency. Intel on the other hand has set sustainability goals and developed metrics for reducing waste (Harmon and Demirkan, 2011). In essence, *green IT* alone does not lead to CR, but may contribute to CR strategies aimed at for instance resource conservation. An example of this is energy conservation. When an organisation's CR target is to increase energy efficiency of business operations, an IT department can contribute. The Chief Information Officer (CIO) could for instance ensure that the IT department would procure more efficient IT equipment or increase the energy efficiency of the already existing infrastructure. *We observe that all strategies tend to lead to more sustainable organizations, but all strategies need support in quantifying, and therewith justifying, the effects of measures a CIO might take. In order to understand the effectiveness of such measures, we will elabor*ate further on how green IT could be measured in the next section.

7.3   MEASURING GREEN IT

This section includes a detailed overview of performance indicators by the sub-parts of *green IT* (Fig. 7.2). Before defining various performance indicators related to *green IT*, performance assessment will be elaborated on briefly. This is important in order to understand how to structure and analyse 'greenness' and how the 'greenness' of IT can be measured.

7.3.1   *Performance assessment*

Green or 'greennes', is a fuzzy concept that is difficult to define because it is dynamic and evolves. A performance indicator is a static measure of an aspect of a criterion that evaluates a particular system performance or output (McDonald and Lane, 2004; Pendlebury *et al.*, 1994; Stein *et al.*, 1994). Each performance indicator may address one or more aspects related to greening IT, such as energy or water use. Such aspects could be defined as an assessment criterion or performance indicator (PI). Assessment criteria are used for comparison and judgement and PIs are used to indicate how well an organisation is doing with respect to predefined targets. Ideally, PIs or assessment criteria defined to measure green IT would be (Custance and Hillier, 2002; Jasch, 2000; 2009; McCool and Stankey, 2001; Nicis Institute, 2010):

- Repeatedly measurable in time
- Target oriented
- Comprehensible, i.e. understandable by non-scientists and other users
- Comparable, i.e. reflect changes in environmental performance
- Compatible, i.e. derived by the same criteria and be related to each other

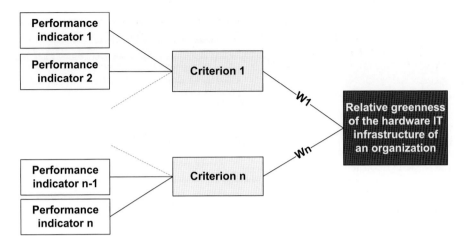

Figure 7.1.   Evaluation system of *green IT* inspired by Zhu *et al.* (2012).

- Balanced, i.e. reflecting environmental performance in a concise way and display problem areas in addition to benefits

Figure 7.1 shows the relationship between performance indicators (PIs) and assessment criteria. It also shows how assessment criteria could be aggregated to determine one value expressing the relative greenness of the hardware IT infrastructure of an organization. To determine this weights can be assigned to the assessment criteria or aggregated PIs to create a composite index (W1, W2, ..., Wn). With regards to weighing assessment criteria, there is no generally accepted procedure in sustainability literature (Bohringer and Jochem, 2007). An option could be to consult experts in an open discussion process or to assign each assessment criterion or PI an equal weight (Bohringer and Jochem, 2007). To aggregate the assessment criteria or PIs scores need to be normalized. Due to lack of a reference system, normalization was not possible in the case study presented in Section 7.6. Before presenting the case study results, performance indicators that may be used to describe *green IT* will be further elaborated on.

7.3.2   *Performance indicators*

Several performance indicators from literature can be used as a measure of IT greenness. Performance indicators from literature are summarized in Figure 7.2. These have been used to design a conceptual framework. The conceptual framework subsequently has been formatively validated by two separate expert panels including academic subject matter experts as well as practitioners.

Figure 7.2 shows several performance indicators related to environmental and economic sustainability, eco-efficiency and greenness of data centers and IT service centers. Environmental and economic performance indicators related to sustainability are derived from international standards that are relatively mature; European Union Eco-management and audit scheme (EMAS), ISO 14031 environmental performance evaluation standard, Global Reporting Initiative (GRI) core set of indicators and Lowell Centre for Sustainable Production (LCSP). The EU environmental management and audit scheme has been available for participation by companies since 1995. The ISO 14031 standard was published in 1999 and has been recommended by the European Commission as part of the EMAS selection and use of environmental performance indicators (European Commission, 2012). Particularly important are the GRI's indicators for sustainability as these are widely used by organizations today in corporate responsibility reports (Roca and Searcy, 2012). The core indicators suggested by LCSP are developed from widely known indicator sets such as the GRI, ISO 14031, World Business Council for Sustainable Development (WBCSD) and centre for waste reduction technologies (Veleva and Ellenbecker, 2001a; 2001b). Despite that

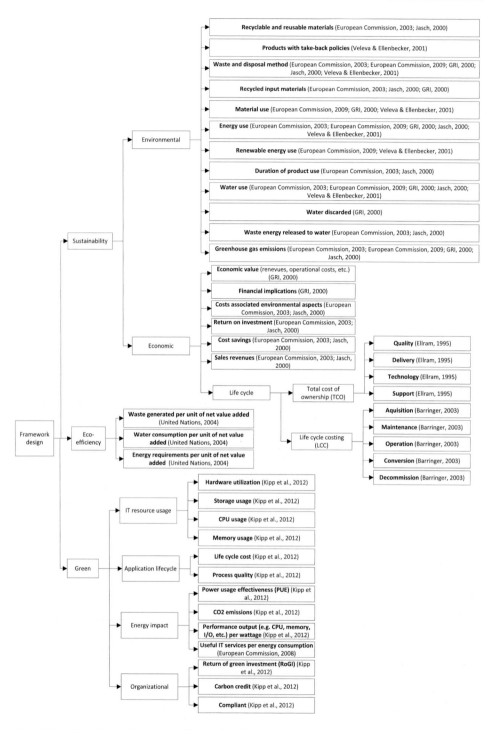

Figure 7.2.   Overview performance indicators from literature.

these environmental performance indicators are internationally recognized, they are not directly applicable due to the specific purpose of the framework. More related indicators are the Green Performance Indicators (GPIs) developed specifically for data centres and IT services centres. These focus mainly on energy use, CO_2 emissions and costs. With regards to eco-efficiency indicators, these might be misleading; if economic value increase faster than environmental impact decrease, the development might seem more environmentally and economically sustainable whereas this is false (Mickwitz, *et al.*, 2006).

The performance indicators shown in Figure 7.2 are used to measure the greenness of IT and embedded in an assessment tool. Prior to conducting an assessment it is important to define the requirements of an assessment tool. This is detailed in the next section.

7.4 DESIGN REQUIREMENTS

The environmental context analysis is essential to ensure practical relevance (Hevner, 2007). The purpose of the environmental context analysis is to define requirements of an assessment tool with the objective of assessing the greenness of an organization's hardware IT infrastructure. These requirements are referred to as design requirements. To define design requirements, core activities of requirements engineering have been used to structure and guide the process. Requirements engineering is applied because it is a structured way of analyzing stakeholder requirements from a systems engineering perspective, which contributes to rigorous requirements (Kotonya and Sommerville, 1988).

The requirements have been elicited through semi-structured interviews and a brainstorm session with relevant stakeholders and experts and from literature. Design requirements have been defined from the stakeholder categories; IT decision-maker, green IT advisor and developer (IEEE, 1998). After elicitation, design requirements have been analysed and developed according to the standard construct. The construct validity of the design requirements have been evaluated by two expert panels. The expert panel evaluation also resulted in a list of core requirements. These serve as a statement of design evaluation that is stable in presence of change and flexible enough to be customized and adapted to changing requirement (Nuseibeh and Easterbrook, 2000).

Requirements for a tool to assess the greenness of the hardware IT infrastructure of an organization are defined in Table 7.2. To develop the assessment tool over time it is important the tool meet the set of core requirements. The main goal of the tool is defined as; *to provide IT decision-makers with an assessment tool to evaluate the greenness of the hardware IT infrastructure of an organization as a step towards a comprehensive CR strategy.*

The core design requirements as well as performance indicators were used to design a framework for green hardware IT infrastructure. In the next section we will describe and present the framework that can be used to measure the greenness of IT.

Table 7.2. Core design requirements.

#	Core design requirements
1	The tool shall determine one value stating the relative greenness of the hardware IT infrastructure of an organization
2	The tool shall demonstrate trade-offs organizations are confronted with in greening the hardware IT infrastructure
3	The tool shall determine the influence of main energy use, water use, raw materials use, CO_2 emission and cost drivers on the relative greenness of the hardware IT infrastructure of an organization
4	The tool shall support organizations improve the relative greenness of the hardware IT infrastructure over time as part of a continuous management process
5	The tool shall produce results that can be compared to peers
6	The tool shall be based upon information from accepted standards
7	The tool shall communicate the results clearly to IT decision-makers

7.5   FRAMEWORK DESIGN

As the essence of design science lays in the scientific evaluation of artefact; the process of designing a new framework focused on formative validation (Livari, 2007). Two expert panels and one case study were conducted to design the framework. In the next sections the new framework will be presented. Preliminary to this, the framework entity will be defined to understand the scope of a comprehensive assessment.

7.5.1   *Framework entity*

The framework entity, which is the object of analysis, is the hardware IT infrastructure of 'an organization'. The adopted definition of an organization is "any coherent entity or activity that transfers a set of inputs into one or more outputs" (Gerbens-Leens and Hoeksta, 2008: 13). This is based on the business or operation of an organization assuming it could be defined by its products (Glenn and Malott, 2004). To assess the relative greenness of the hardware IT infrastructure of an organization, the organization has to be delineated from its environment. The framework entity is inspired by two information sources: the ISO 14031 standard for environmental performance evaluation and the GHG Protocol. The GHG Protocol is the most widely used standard (Kolka *et al.*, 2008). The next sections describe the integration of these sources into a comprehensive framework entity design that describes an organization's hardware IT infrastructure supporting business applications. The framework entity is presented in Figure 7.3.

The high-level "building blocks" of an organization are inspired by the ISO 14031 standard for environmental performance evaluation. In the ISO 14031 standard an organization is defined according to its operation. It is defined as consisting of physical facilities and equipment as well as supply and delivery from them (Jasch, 2000). The life cycle of hardware IT incorporated in this study is limited to procure (supply), use and dispose (delivery).

The procurement stage begins when the acquisition process starts and ends when the acquisition is done and including installing the equipment so that it might be used. Derived from the consumer responsibility principle the following might be defined; when an organization is the first buyer of a hardware IT asset, the environmental and economic impacts of the preliminary life cycle stages of the asset fall under the procurement stage and respectively the responsibility of the organization (Lenzen *et al.*, 2007). A *hardware IT asset* is defined as an outsourced, leased and/or bought hardware IT infrastructure unit (WRI and WBSCD, 2011).

The use stage starts when the consuming organization takes possession of the product and ends when the product is discarded to a waste treatment location (WRI and WBSCD, 2011). The use of hardware IT consist of the use of owned, outsourced and leased hardware IT infrastructure units. This is in accordance with the GHG protocol scope 3. Outsourced activities are included when these support an organization's business applications, such as an outsourced data centre or

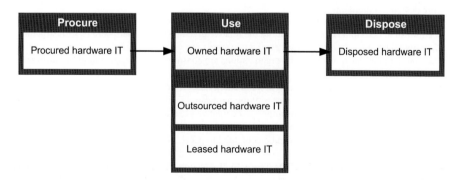

Figure 7.3.   Framework entity design.

IT service centre (WBCSD and WRI, 2004). With respect to leased assets, an organization is only accountable for those that it operates (WBCSD and WRI, 2011a).

Disposal or end-of-life starts when used products are discarded by the organization and ends when the product is returned to nature or allocated to another product's life cycle. This includes the environmental effect of third-party disposal and treatment of waste generated by reporting organization's owned or controlled operations (WBCSD and WRI, 2011b). Hence, an organization is only responsible for the disposal of owned hardware IT assets (WBCSD and WRI, 2004).

For the purpose of this research the units of analysis are defined as individual hardware IT infrastructure units such as a laptop type HP Folio 13.

7.5.2   Framework design

In the previous sections, the framework entity has been defined. In Table 7.3 the new framework, entitled the *green hardware IT Infrastructure (GHITI) framework* is presented. The framework consists of several performance indicators related to resource usage (energy, water and raw materials), GHG emissions, waste electronic and electric equipment (WEEE) and costs related to hardware IT supporting business applications.

On the rows there are five classes of environmental and economic assessment criteria and PIs. Each of these address one generic issue an organization is confronted with when greening the hardware IT infrastructure: water use, energy use, raw materials waste generation, GHG emission and costs. The environmental assessment criteria are chosen on the vertical axis of the framework because the aggregation of performance indicators that fall under these result in five values that make trade-offs easier to comprehend and communicate to IT decision-makers (see design requirements).

7.6   CASE STUDY

As part of evaluating the framework, a case study has been conducted. The objective of the case study was to evaluate the viability of the performance indicators. The case was limited to the Dutch part of an international organisation. The organization has more than 3000 employees in The Netherlands distributed over 10 offices. Further, the organisation has an extensive hardware IT infrastructure supporting business applications. Approximately 80 units of analysis found within the organization were incorporated in the case study. In the following sub-sections the research scope, data sources, data collection procedures, data analysis and evaluation will be elaborated on.

7.6.1   Research scope

The scope of the case study is limited to the hardware IT infrastructure used in offices and data centres at the organization's head-office in The Netherlands. The organization does not lease hardware IT, however the operation of several units is outsourced to two data centre vendors. The hardware IT infrastructure units in these data centres are included in the analysis. In this research the hardware IT infrastructure units that are listed in the *asset list* were included. It is recognized that next to these units there might be other hardware IT infrastructure units in use at the organization investigated at the moment of performing the research. End-user equipment has for instance been delineated to peripherals and computers, whereas it is recognized employees use smart-phones, monitors, etc. The classification of units of analysis incorporated in the research scope is shown in Figure 7.4.

Within each class of hardware IT infrastructure units defined in Figure 7.4, units of analysis are defined such as HP EliteBook 8460p: laptop. The classification adopted has been defined with input from *KPMG IT Advisory* and *CIO Platform*[1] and literature (Beccalli, 2007; Gurbaxani *et al.*, 1998; ITL Eduction Solutions, 2006; Unhelkar, 2011).

---

[1] In November 2011 participants at the CIO Platform in The Netherlands (http://www.cio-platform.nl/) signed an IT-energy efficiency covenant. The presented framework was discussed in workshops with representatives of member organizations of the CIO Platform. The workshops were facilitated by KPMG IT Advisory.

Table 7.3. Green hardware IT Infrastructure (GHITI) framework.

Assessment criteria	Life cycle phase		
	Procure performance indicators	Use performance indicators	Dispose performance indicators
1. Water use over the life cycle of hardware IT (m^3)	1.1.1 Amount of water use related to the extraction and production process of hardware IT	1.2.1 Amount of water used in hardware IT operation per water source	1.3.1 Amount of water use related to hardware IT discarded to landfill 1.3.2 Amount of water use related to recycling hardware IT 1.3.3 Amount of water use related to recovering energy from hardware IT 1.3.4 Amount of water use related to reuse and refurbish hardware IT for third party use
2. Energy use over the life cycle of hardware IT (MJ)	2.1.1 Amount of energy use related to the extraction and production process of hardware IT	2.2.1 Amount of energy use of hardware IT operation 2.2.2 Utilization of hardware IT relative to unit specific optima (%)	2.3.1 Amount of energy use related to hardware IT discarded to landfill 2.3.2 Amount of energy use related to recycling hardware IT 2.3.3 Amount of energy recovered from discarded hardware IT 2.3.4 Amount of energy use related to reuse and refurbish hardware IT for third party use
3. Raw material waste generation over the life cycle of hardware IT (kg)	3.1.1 Amount of recycled and reused materials in hardware IT procured 3.1.2 Amount of raw materials in hardware IT procured 3.1.3 Amount of refurbished or used hardware IT procured	3.2.1 Amount of replaced sub-components in hardware IT 3.2.2 Amount of repaired sub-components in hardware IT	3.3.1 Amount of materials from hardware IT to landfill 3.3.2 Amount of recycled materials from hardware IT 3.3.3 Amount of hardware IT discarded for energy recovery 3.3.4 Amount of materials in hardware IT reused and refurbished for third party use
4. Greenhouse gas emissions over the life cycle of hardware IT (ton CO$_2$ equivalent)	4.1.1 Amount of greenhouse gas emission related to the extraction and production process of hardware IT	4.2.1 Amount of greenhouse gas emission related to extraction and production of energy used in hardware IT operation	4.3.1 Amount of greenhouse gas emission related to hardware IT discarded to landfill 4.3.2 Amount of greenhouse gas emission related to recycling hardware IT 4.3.3 Amount of greenhouse gas emission related to recovering energy from hardware IT 4.3.4 Amount of greenhouse gas emission related to reuse and refurbish hardware IT for third party use
5. Costs over the life cycle of hardware IT (euro)	5.1.1 Procurement costs of hardware IT	5.2.1 Energy costs of hardware IT operation 5.2.2 Water costs of hardware IT operation 5.2.3 Costs of carbon credits for the operation of hardware IT 5.2.4 Maintenance costs of hardware IT	5.3.1 Costs of hardware IT discarded to landfill 5.3.2 Costs of recycling hardware IT 5.3.3 Costs of recovering energy from discarded hardware IT 5.3.4 Revenues from selling used and refurbished hardware IT to third parties

7.6.2 *Data sources and data collection procedure*

Data was collected to estimate several performance indicators. To estimate the greenness of the hardware IT infrastructure defined in Figure 7.4, various data sources were used to ensure methodological triangulation (Yin, 2009); documentations, archival records and interviews. To ensure data source triangulation, interviews as well as separate documentation were asked for ver-ification of data collected external to the organization investigated. The process was streamlined to ensure efficient collection of data. Data collected was verified by IT managers. Nevertheless several assumptions were made due to lack of accessible or short-termed availability of data. Per-formance indicators related to landfill and energy recovery from waste, repaired sub-components,

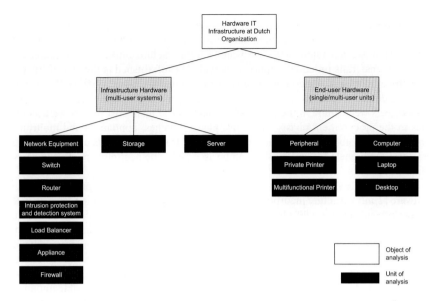

Figure 7.4.   Categories of units of analysis in case study scope.

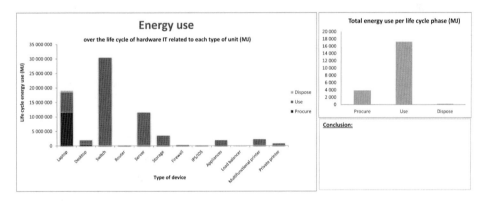

Figure 7.5.   Energy use over the life cycle of hardware IT at the organization investigated.

end-of-life costs and water costs were assumed to be negligible. All other performance indicators in Table 7.3 were incorporated in the case study research. The analysis of the data and the results will be presented in the next section.

7.6.3   *Data analysis and evaluation*

In the previous sections the goal, scope and data collection procedure have been described. In this section the results of the analysis of the data will be presented. In the subsequent sections the results will be presented per assessment criteria or PI as defined in the GHITI Framework (Table 7.3).

Figure 7.5 shows that the main energy footprint drivers are switches and servers. Laptops are the only device where energy use in the procurement phase exceeds the use phase. For each PI we can generate a similar overview. We present a brief overview of the main conclusions of the case before we reflect on the findings.

Most water is used during the procurement phase and is related to extraction of raw materials and production. For each assessment criterion or PI the results can be presented this way. This allows us to see which units of hardware cause the greatest environmental impact. The main driver of the water use over the life cycle of assessed hardware IT is laptops. The largest amount of water use can be found in the procurement phase, hence production of hardware unit. Looking at raw material waste generated by the hardware IT infrastructure units assessed, almost the same amount of raw materials procured (input) are recycled or refurbished at the end-of-life (output). The main driver of the raw material waste generation is multifunctional printers. Waste generated by replacing defective devices in the use phase was found to be negligible in the case study. The largest costs were found in the procurement phase. The main cost drivers were storage and switches in the procurement phase. In the use phase switches, appliances and multifunctional printers were responsible for the highest costs. Income from selling units at the end-of-life (i.e. disposal) was the highest for laptops, but can be assumed to be negligible relative to costs made in preliminary phases.

The estimates made show areas that the organisation could address to improve the greenness of the hardware IT infrastructure incorporated in the case study. The organisation investigated could for instance procure laptops with a lower water footprint or more energy efficient switches in order to meet targets set at lowering the water and energy footprint of the estimated IT infrastructure. By procuring second hand laptops, water use over the life cycle of hardware IT could be lowered by approximately 6 million m^3. In a technology driven organisation as the company investigated, procuring second hand IT was not an option. However, procuring more energy efficient switches or incorporating water footprint criteria in procurement policies could be possible initiatives. The framework will be evaluated in the next section.

7.7   EVALUATION AND REFLECTION

In the previous section the implementation of the new framework has been described. The next step is to evaluate the framework design. In the following sections the framework will be evaluated and reflected on to understand its advantages, limitations and potential improvements.

7.7.1   *Evaluation of the framework*

The core set of design requirements defined in Section 7.6 is stable enough to be used to evaluate the framework, at the same time it is important and necessary to develop a tool that is flexible enough to adapt changing requirements (Nuseibeh and Easterbrook, 2000). The framework design, as implemented in a single, embedded case study has shown to fulfil most of these requirements. The case study shows that;

- Performance indicators may be measured and aggregated into one value stating the relative greenness of the hardware IT infrastructure investigated;
  - A single value facilitates decision making in organizations. Green IT is important but not on the top of mind of most decision makers. Multiple values will result in an increased complexity of decision making. This will not contribute to achieving targets set for greening IT such as increasing energy efficiency.
- Main drivers may be identified that cause the level of greenness of the hardware IT infrastructure investigated;
  - To ensure a rigorous implementation of measures it is essential to identify the main drivers first. Low hanging fruit will increase the adoption of required measures and contribute to achieving the target set for greening IT. The case study has for example shown that procuring more energy efficient switches might have a significant effect on energy consumption.
- The trade-offs the organization are confronted with when greening the hardware IT infrastructure are evident when applying the framework;
  - Trade-offs in the investigated case study were predominantly trade-offs between costs and energy use of IT. Facilitating the decision making by understanding the implications related to energy efficiency is helpful. This may also provide insights about potential reductions in CO_2 emissions caused by IT-energy efficiency initiatives.

- The framework may support an organization improve the relative greenness of the hardware IT infrastructure over time if incorporated in a continuous management process (see Section 7.8);
  - Improving the greenness of IT is not a one-off implementation of a framework. The highest value can be achieved when the framework is part of a continuous management process that continuously increases the greenness of IT over time. The framework facilitates embedding IT-energy efficiency, GHG emissions, water consumptions, generation of IT-waste and energy use in the decision making process.
- Assessment criteria scores that are defined per employee or in a financial fashion can be used to compare results with peers;
  - Benchmarking criteria scores increases management attention for water use, energy use and efficiency, GHG emissions and waste that is caused by the IT infrastructure. Awareness of the IT-energy efficiency performance compared to peers can facilitate the decision making process. However it is important to understand the specific characteristics of organizations if scores are compared.
- Although new in its sort, the framework is based upon accepted business standards related to environmental sustainability and IT investment evaluation. A case study shows this does not necessarily imply all data are readily available or accessible from within the organization;
  - When implementing the framework it was found to be more important to start measuring than to focus on a full measurement of the entire IT infrastructure. Overtime additional performance indicators can be added to the framework measurement.
- When presentation layouts are carefully designed and tailored to the needs of an IT decision-maker, results may be communicated more effectively;
  - Spending time with stakeholders on the presentation of the management report is time well spent. Furthermore a timely evaluation of the management report is important and may result in adjustments in the report that enhances understanding of results.

The evaluation of the design requirements shows that implementing the framework design together with the recommended implementation may lead to sustained use of the framework at the organization investigated. Although the framework meets the core design requirements quite well in the first evaluation, it has several limitations. In the next sections the limitations of the new framework will be elaborated to better understand its applicability and areas of improvement.

7.7.2   Reflection on the framework

As suggested in the previous sections, using the framework over time as part of a continuous management process contributes to the CR strategy of an organization. With regards to environmental sustainability, the framework addresses several issues; energy use, water use, raw materials, generation of electronic and electrical waste and CO_2 emission. Improving the framework would imply more issues are incorporated such as land use, biodiversity, water pollution, emission of ozone depletion substances, NO_x. SO_x, and other air emissions (GRI, 2000b). Addressing economic sustainability, the framework is narrowed down to costs. This may be improved by incorporating other economic sustainability aspects related to quality, market presence, indirect economic impacts, openness to stakeholder review and participation in decision-making process (GRI, 2000a; Veleva and Ellenbecker, 2001a).

Besides extending the scope of both economic and environmental sustainability topics, the main improvement potential of the framework may be found in the social dimension of sustainability. Despite that social aspects of hardware IT have not been incorporated in the research scope, we recognize the importance of ensuring comprehensive improvements of the current hardware IT industry, starting at organizational level, that also includes aspects related to the well-being of humans. Nevertheless, it is recognized that economic, social and environmental sustainability are closely intertwined and that in practice they may be difficult to separate (Rao and Holt, 2005). Thus, environmental and economic aspects incorporated in the framework will have an effect on the social sustainability dimension as well. This does not however imply the framework balance the 3BL. It is recognized that social sustainability is important in obtaining an improved balance of the 3BL and thereby also to a comprehensive CR strategy. Only when all three dimensions of sustainability are addressed in a balanced way and when the needs of the present does not comprise the ability of future generations to meet their needs, it may be entitled sustainable development and a step towards a balanced, comprehensive CR (Brundtland, 1987; Labuschagne *et al.*, 2005).

Currently the infrastructure, when implemented, is expected to contribute to cost savings, through energy, software and hardware reductions. Thus a substantial part of the effort where eco-efficiency efforts are concerned can be made visible. But how do we come to implementation? This will be the subject of the following section.

7.8   PRACTICAL RECOMMENDATIONS ON IMPLEMENTATION

In the previous section the framework has been evaluated. In this section a translation is made from design to implementation. We do so by suggesting a strategy to deploy the framework design. This entails a guideline for further application of the functional design and a vision on how to manage the process of measuring the greenness of hardware IT. The vision has been based upon insights from the case study as well as literature on environmental performance evaluation.

7.8.1   *Applied principles of functional design*

Based on the results from the evaluation in Section 7.7 we formulated guidelines should be adhered to when applying the functional design in its current form. Not only is a proper functional design important, attention to implementation is also relevant and will be dealt with in Subsection 7.8.2.

1. *Keep it up-to-date*
   Due to stated limitations of the case study protocol in its current form, it is recommended this is improved to enhance quality of data collected. Despite that it is recommended to improve the case study protocol, it should be used consistently in additional business cases to ensure reliability of data and comparability of results for internal as well as external benchmarking purposes. This way, additional business cases can build on earlier successes.
2. *Tailor presentation layout*
   Presentation layouts have been suggested in Section 7.7. However, these should be tailored to its audience to increase the chances that the information presented leads to behaviour that supports the organization.
3. *Use the available and accessible information*
   Information needed to conduct calculations has to be available and accessible within organizations to perform reliable estimations of performance indicators. This would require passive and active monitoring of data. Where data cannot be found in financial statements or assets lists, data might be found through interviews with IT managers or experts, hardware IT suppliers and end-of-life vendors. Reference data might also be found in literature. Without required data being available, the functional design cannot be applied successfully.
4. *User should have basic knowledge*
   It is recommended the user of the framework has basic knowledge of IT and green IT assessment. The framework and functional design provide guidance on how to assess the relative greenness of the hardware IT infrastructure of an organization, but still requires some level of understanding of the topic by the user of the framework. Additionally, the user should be familiar with data analysis software such as Excel. Such software is used to calculate and aggregate performance indicators.
5. *Communicate and extend the framework use*
   It is recommended to extend the scope of the hardware IT infrastructure units that can be estimated using the functional design. Possibly calculation methodologies need to be adjusted for a specific type of unit. In this case the changes should be implemented in the functional design consistently. The changes should also be communicated.

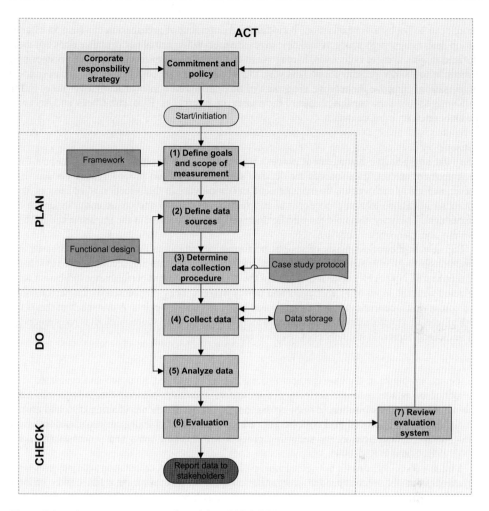

Figure 7.6.   Management process, adopted from ISO 14031.

The next section describes a vision on how to manage the process of measuring the greenness of hardware IT. It is useful to understand how the framework can be applied and how the measurement process can be improved continuously.

7.8.2   *Vision on management process*

It is recommended to make the GHITI framework an integral part of an organization's environmental performance assessment. The ISO 14000 environmental management standard describes a process for measuring environmental performance. This is based upon the Plan-Do-Check-Act business process improvement model. In line with the ISO 14031 standard, a management cycle is recommended to measure and improve the greenness of the hardware IT infrastructure of an organization over time. Below we present such a cycle to support management in their efforts to increase the environmental performance of the organization. This process is depicted in Figure 7.6.

Figure 7.6 shows how an organisation can measure the greenness of the hardware IT infrastructure of an organisation and improve the measurement process over time. Information from this process can also be used to improve performance. Targets and objectives regarding GHITI scores can be set by an organisation. Periodical measurement of progression can help an organisation determine whether targets and objectives are met. When actual performance does not meet defined targets to meet objectives, management can choose to implement measures to steer performance in the desired direction. This way, the management process in Figure 7.6 can support organisations improve the relative greenness of the hardware IT infrastructure over time. In the following sections the various phases of the management process (Plan, Do, Check and Act) will be elaborated on in more detail.

Plan

The first phase in the management process is planning. In this phase goals and scope of the measurement are set, data sources are defined and data collection procedures are chosen. It is recommended to involve middle management and employees in the selection of indicators, data sources and data collection procedures to ensure data availability and create commitment to the assessment. This may also hold the middle management accountable at the implementation stage (Veleva and Ellenbecker, 2001a).

The new framework consists of several performance indicators that can be complicated and time-consuming to measure in a real business case. To overcome this, it is recommended to select a limited number of performance indicators. In the case study relevant performance indicators were selected with the input of IT managers and policy makers. They pinpointed which performance indicators would be irrelevant or negligible and where to collect the data. A drawback of this is the risk of selecting performance indicators on which the organisation scores high. It is recommended to think long-term and not only choose quick wins.

Do

In the second phase in the management process, data is collected, analysed and evaluated. The collection of data to estimate performance indicators should be planned carefully to minimize the burden on an organisation, risk of errors that might occur when collecting data and ensure data collected is approved and collected on a consistent basis. Ideally, an organisation would integrate the data collection and reporting process with existing tools and processes (Putnam, 2002; WBCS and WRI, 2004).

A trade-off should be made between detail level of data collected, calculation accuracy, available time and budget. It is recommended an organisation apply the most accurate, reliable and verifiable data. Completeness of data is also necessary to ensure a comprehensive analysis of performance. Enforcing data that meet all these criteria might be difficult due to time and budget constraints. In the case study the entire data sheet was sent to IT managers for verification. This is recommended as part of data collection procedures to enhance data quality. An external party could verify the reliability of this data as well to further enhance the quality. Next to the importance of high data quality, it is recommended to streamline data collection processes to ensure data is collection efficiently. Also cross checking collected data will improve the data quality.

Regarding the analysis of data it is recommended to use the functional design consistently. This way results can be used for internal and external benchmarking. Benchmarking can for instance be used to compare the GHITI performance with organisation units in other countries. Ideally one compares similar units, but the legal unit might be different in for example size and hardware IT infrastructure. Further, it is recommended to measure the same classes of units of analysis. This way the resulting scores per unit class can be compared although units of analysis change over time.

Evaluation of an organisation's performance is the last step in this phase. The greenness of the hardware IT infrastructure should be reviewed periodically. A consistent and meaningful comparison of the GHITI scores over time also requires that organisations set a performance datum to compare the current GHITI level. The performance datum can be referred to as the base year (WBCS and WRI, 2004). To consistently track the GHITI scores over time, the GHITI scores in the base year might have to be recalculated as organisations undergo significant structural changes such as mergers, acquisitions and divestments, outsourcing or in sourcing of activities and the implementation of shared service centers influencing the greenness level of the hardware IT. The threshold for recalculating the base year is dependent on the significance of the change (WBCS and WRI, 2004). An organisation could use the base year as a basis for setting and tracking progress towards a "GHITI score" objective at a specific point in time.

Check
In this phase of the management process the measured performance is communicated and the GHITI evaluation is reviewed and improved. It is recommended to communicate the results to internal and external stakeholders to create awareness, demonstrate commitment and put informa-tion in the hands of actors responsible for improvements (Putnam, 2002). Besides communicating the results, it is essential to improve the GHITI evaluation over time.

Act
The CR strategy is used as input to define goals and scope of the measurement in line with the framework. Commitments and policies in the CR strategy trigger the implementation of an assessment framework for green IT. In the case study organization the CR strategy will be linked to IT-energy efficiency. In the future the Corporate Social Responsibility (CSR) report of the case study organization will include IT-energy efficiency metrics.

Furthermore performance indicators incorporated in the framework should be evaluated to achieve a continuous improvement of the functional design and the evaluation process (Veleva and Ellenbecker, 2001a).

7.9   CONCLUSION AND FURTHER RESEARCH

We conclude that the framework presented in Table 7.3 can be used to determine the relative greenness of the hardware IT infrastructure of an organization as a step towards a comprehensive CR strategy. Secondly, measuring progression in *greenness* requires that the framework becomes embedded in a continuous measurement and management process. This way, the framework can support achieving the desired *green IT* performance indicators scores and long-term goals. Organizations can use the outcome of periodical measurements from the framework to, if required, adjust their policies in order to achieve the CR goals as part of their CR strategy. The framework incorporates economic and environmental performance indicators derived from several accepted standards. The framework can be used to assess *green IT* progression related to energy use and efficiency, water use, generation of raw material waste, GHG emission and costs over the life cycle of hardware IT.

Furthermore, it is recommended additional design cycles are implemented to further verify, validate and refine the framework design. It could be particularly useful to perform additional case studies to refine the functional design and test the general applicability of the new framework in terms of units of analysis. The validity of the output of the assessment tool should be tested as well.

As outlined in the evaluation and reflection, completeness is not claimed. To further broaden the framework in terms of sustainability and CR strategy, the framework should also incorporate a social dimension. The upcoming research field of social life cycle analysis (LCA) might constitute a valuable input for this. During the research it was encountered that LCA data often was very limited. No data could be found about the water and energy use related to materials extraction and production of switches, routers, firewalls, servers, storage, appliances, etc. Data could be obtained from hardware IT suppliers; however the reliability of this is questionable unless verified by a third party. Due to this, it is recommended to investigate the environmental impacts of these types of units more in-depth from a scientific perspective. This could improve data quality and the accuracy of the functional design. It could possibly also fill a knowledge gap found in literature on LCA. Additional environmental and economic issues could be incorporated in an extension of the new framework as well.

The authors would like to thank interviewees and experts for sharing their valuable insights. In addition the internship of Johanne was of great value for writing this chapter. The authors would like to thank the company for facilitating Johanne's internship and supporting her thesis research. Finally the authors would like to thank the reviewers for their constructive feedback. Their feedback has improved the readability of this chapter.

Ir. Johanne Punte Kalsheim is a graduate student from Delft University of Technology, The Netherlands and employed by Ernst & Young in Norway.

Prof. dr. Erik Beulen holds the Global Sourcing chair at Tilburg University, The Netherlands and is employed by KPMG.

REFERENCES

Beccalli, E.: Does IT investment improve bank performance? Evidence from Europe. *Journal of Banking and Finance* 31:7 (2007), pp. 2205–2230.

Bohringer, C. & Jochem, P.E.: Measuring the immeasurable – a survey of sustainability indices. *Ecological Economics* 63:1 (2007), pp. 1–8.

Broadbent, M. & Weil, P.: Management by maxim: how business and it managers can create it infrastructures. *Sloan Management Review* 38 (1997), pp. 77–92.

Brundtland, C.: Our common future. Vol 383. *United Nations World Commission on Environment and Development*, Oxford University Press, Oxford, UK, 1987.

Capra, E. & Merlo, F.: Green IT: everything starts from the software. *Proceedings of the European Conference on Information Systems*, Verona, Italy, 2009, pp. 62–73.

Cone, E.: The greening of the CIO. *CIO Insights*, posted November 7, 2006.

Cone. (2006). The greening of the CIO. CIO insights.

Custance, J. & Hillier, H.: Statistical issues in developing indicators of sustainable development. *Journal of the Royal Statistical Society*: Series A (*Statistics in Society*) 161:3 (2002), pp. 281–290.

Da Silva, L.F. & e Abreu, F.B.: Reengineering it infrastructures: a method for topology discovery. 7th International *Conference on the Quality of Information and Communications Technology, QUATIC*, 2010, pp. 331–336.

Dao, V., Langella, I. & Carbo, J.: From green to sustainability: information technology and an integrated sustainability framework. *Journal of Strategic Information Systems* 20:1 (2011), pp. 63–79.

Duncan, N.B.: Capturing flexibility of information technology infrastructure: a study of resource characteristics and their measure. *Journal of Management Information Systems* 11:3 (1995), pp. 37–57.

Elliot, S.: Transdisciplinary perspectives on environmental sustainability: a resource base and framework for IT-enabled business transformation. *MIS Quarterly* 35:1 (2011), pp. 197–236.

Ellram, L.M.: Total cost of ownership: an analysis approach for purchasing. *International Journal of Physical Distribution & Logistics* 25:8 (1995), pp. 4–23.

Erek, K., Loeser, F., Schmidt, H.H., Zarnekow, R. & Kolbe, L.M.: Green IT strategies: a case study based framework for aligning green IT with competitive environmental strategies. *PACIS 2011 Proceedings*, Association of Information Systems, Brisbane, Australia, 2011.

European Commission: Environement: overview of the take-up of EMAS accross the years, 2012, http://ec.europa.eu/environment/emas/documents/articles_en.htm (accessed June 2012).

Gallego, I.: The use of economic, social and environmental indicators as a measure of sustainable development in Spain. *Corporate Social Responsibility and Environmental Management* 13:2 (2006), pp. 78–97.

Gerbens-Leenes, P. & Hoekstra, A.: Business water footprint accounting: a tool to assess how production of goods and services impacts on freshwater resources worldwide. *Value of Water Report Series* No. 27 UNESCO-IHE, Institute for Water Education, Delft, The Netherlands, 2008.

Glenn, S.S. & Malott, M.E.: Lead article complexity and selection: implications for organizational change. *Behavior and Social Issues* 13: (2004), pp. 89–106.

GRI (2000a) Indicator set economic (EC) version 3.1. Retrieved January 30, 2012, from GRI: http://www.globalreporting.org (accessed January 2012).

GRI (2000b) Indicator set society (SO) version 3.1. Retrieved January 30, 2012, from GRI: http://www.globalreporting.org (accessed January 2012).

Gurbaxani, V. Melville, N. & Kraemer, K.: Disaggregating the return on investment to it capital. *International Conference on Computers and Information Systems*, Association for Information Systems, Atlanta, GA, 1998, pp. 376–380.

Harmon, R.R. & Auseklis, N.: Sustainable IT services: assessing the impact of green computing practices. *Portland International Conference on Management, Engnieering & Technology* (*PICMET*), 2009, pp. 1707–1717.

Harmon, R.R. & Demirkan, H.: The next wave of sustainable IT. *IT Professional* 13:1 IEEE Computer Society, 2011, pp. 19–25.

Harmon, R.R., Demirkan, H. Auseklis, N. & Reinoso, M.: From green computing to sustainable IT: developing a sustainable service orientation. *Proceedings of the 43rd Hawaii International Conference on System Science*, Hawai, 2010, pp. 1–10.

Hevner, A.R., March, S.T., Park, J. & Ram, S.: Design science in information systems research. *Management Information Systems Quarterly* 28:1 (2004), pp. 75–105.

Hird, G.: Green IT in practice: how one company is approaching the greening of its IT. IT Governance Ltd. Cambridgeshire, UK, 2010.

Gerbens-Leens, P.W. & Hoeksta, & A.Y. *Business water footprint accounting: a tool to assess how production of goods and services impacts on freshwater resources worldwide*. Delft: UNESCO-IHE, 2008.

Huppes, G. & Ishikawa, M.: Eco-efficiency and its terminology. *Journal of Industrial Ecology* 9:4 (2005), pp. 43–46.

IEEE: IEEE recommended practice for software requirements specification. IEEE standard 830-1998, 1998.

ITL Eduction Solutions: *Introduction to computer science*. Pearson Education, Delhi, India, 2006.

Jasch, C.: Environmental performance evaluation and indicators. *Journal of Cleaner Production* 8:1 (2000), pp. 79–88.

Jasch, C.: *Environmental and material flow cost accounting: principles and procedures*. Springer Science, Vienna, Austria, 2009.

Kipp, A. Jiang, T. Fugini, M. & Salomie, I.: Layered green performance indicators. *Future Generation Computer Systems* 28:2 (2012), pp. 478–489.

Labuschagne, C., Brent, A.C. & Erck, R.P.: Assessing the sustainability performance of industries. *Journal of Cleaner Production* 13:4 (2005), pp. 373–385.

Lamb, J.P.: The greening of IT. How companies can make a difference for the environment. IBM/Pearson, Detroit, MI, 2009.

Li, Q. & Zhou, M.: The survey and future evolution of green computing. *IEEE/ACM Conference on Green Computing and Communications (GreenCom)*, Sichuan, China, 2011, pp. 230–233.

Linberg, K.R.: Software developer perceptions about software project failure: a case study. *The Journal of Systems and Software* 49:2 (1999), pp. 177–192.

Linnanen, V. & Panapanaan V.M.: Roadmapping corporate social responsibility in Finnish companies. Helsinki University of Technology, Helsinki, Finland, 2002.

Lo, S.F. & Sheu, H.J.: Is corporate sustainability a value increasing strategy for business? *Corporate Governance: An International Review* 15:2 (2007), pp. 345–358.

Martin, E.W., Brown, C.V., DeHayes, D.W., Hoffer, J.A. & Perkins, W.C.: *Managing information technology*. 6th edn., Prentice Hall, 2009.

McCool, S.F. & Stankey, G.: Representing the future: a framework for evaluating the utility of indicators in the search for sustainable forest management. In: R.J. Raison, A.G. Brown & D.W. Flinn (eds): *Criteria and indicators for sustainable forest management*. CAB International, Wallingford, UK, 2001, pp. 93–109.

McDonald, G.T. & Lane, M.B.: Converging global indicators for sustainable forest management. *Forest Policy and Economics* 6:1 (2004), pp. 63–70.

McDonough, W. & Braungart, M.: Design for the triple top line: new tools for sustainable commerce. *Corporate Environmental Strategy* 9:3 (2002), pp. 251–258.

McWilliams, A. & Siegel, D.: Corporate social responsibility: a theory of the firm perspective. *The Academy of Management Review* 26:1 (2001), pp. 117–127.

Mickwitz, P., Melanen, M., Rosenström, U. & Seppälä, J.: Regional eco-efficiency indicators a participatory approach. *Journal of Cleaner Production* 14:18 (2006), pp. 1603–1611.

Molla, A.: Organizational motivations for green it: exploring green it matrix and motival models. *Proceedings of the Pacific Asia Conference on Information Systems*, Hyderabad, India, 2009a, p. 13.

Molla, A., The reach and richness of Green IT: a principal component analysis: *20th Australasian Conference on Information Systems*, 2009b, Melbourne, Australia, pp. 754–764.

Molla, A. & Cooper, V.: Green IT readiness – a framework and preliminary proof of concept. *Aystakasian Journal of Information Systems* 16:2 (2009), pp. 5–23.

Molla, A., Cooper, V. & Pittayachawan, S.: The green IT readiness (G-readiness) of organizations: an exploratory analysis of a construct and instrument. *Communications of the Association for Information Systems* 29:1 (2011), pp. 67–96.

Murugesan, S.: Harnessing green it: principles and practices. *IT Professional* 10:1 2008, pp. 24–33.

Nicis Institute: Handboek prestatiemeting, Den Haag. Nicis Institute and Delft Universtity of Technology, Delft, The Netherlands, not dated.

Nuseibeh, B. & Easterbrook, S.: Requirements engineering: a roadmap. *Proceedings of the Conference on the Future of Software Engineering*, Edinburgh, UK, 2000, pp. 37–46.

Orsato, R.J.: *Sustainability strategy*. Palgrave Mcmillan, Hampshire, UK, 2009.

Pendlebury, M., Jones, R. & Karbhari, Y.: Developments in the accountability and financial reporting pratices of executive agencies. *Financial Accountability & Management* 10:1 (1994), pp. 33–46.

Porter, M.E. & Kramer, M.R.: Strategy and society: the link between advantage and corporate social responsibility. *Harward Business Review* 84:12 (2006), pp. 78–92.

Rao, P. & Holt, D.: Do green supply chains lead to competitiveness and economic performance? *International Journal of Operations and Production Management* 25:9 (2005), pp. 898–916.

Roca, L.C. & Searcy, C.: An analysis of indicators disclosed in corporate sustainability reports. *Journal of Cleaner Production* 10:1 (2012), pp. 103–117.

SAIC: Life cycle assessment: principle and practices. National Risk Management Research Laboratory, US Environmental Protection Agency, Cincinnati, OH, 2006.

Schaltegger, S. & Wagner, M.: Managing the business case for sustainability. *Proceedings EMAN-EU Conference*, 2008, pp. 7–9.

Skouloudis, A., Evangelinos, K. & Kourmousis, F.: Assessing non-financial reports according to the global reporting initiative guidelines: evidence from Greece. *Journal of Cleaner Production* 18:5 (2010), pp. 426–438.

Steina, A., Rileyb, J. & Halbergc, N.: Issues of scale for environmental indicators. *Agriculture, Ecosystems & Environment* 87:2 (2001), pp. 215–232.

Trusty: An overview of life cycle assessments: part one. *International Code of Council Building Safety Journal Online* (2010), Vol. 7, pp. 1–4.

Unhelkar: *Green IT strategies and applications*. CRC Press, Boca Raton, FL, 2011.

United Nations: A manual for the preparers and users of eco-efficiency indicators, Version 1.1. United Nations, Geneva, Switzerland, 2004.

Veleva, V. & Ellenbecker, M.: Indicators of sustainable production: framework and methodology. *Journal of Cleaner Production* 9:6 (2001a), pp. 519–549.

Veleva, V. & Ellenbecker, M.: A proposal for measuring business sustainability: addressing shortcomings in existing frameworks. *Greener Management International* 31 (2001b), pp. 101–120.

Velte, T., Velte, A. & Elsenpeter, R.: *Green IT – Reducing your information system's environmental impact*. McGraw-Hill, New York, 2008.

Watson, R.T., Boudreau, M.C. & Chen, A.J.: Information systems and environmentally sustainable development: energy informatics and new directions for the IS community. *MIS Quarterly* (2010), pp. 22–38.

WBCSD: Weco-efficiency: creating more value with less impact. World Business Council for Sustainable Development, 2000.

WBCSD & WRI: GHG protocol: a corporate accounting and reporting stanard. World Resources Institute and World Business Council for Sustainable Development, Washington, DC, 2004.

WBCSD & WRI: GHG Protocol Product Life Cycle Accounting and Reporting Standard – ICT Sector Guidance. Washington: World Business Council for Sustainable Development and World Resource Institute, 2012.

WBCSD & WRI: Product life cycle accounting and reporting standard. World Resources Institute and World Business Council for Sustainable Development, Washington, DC, 2011a.

WBCSD & WRI: Corporate value chain (scope 3) accounting and reporting standard. World Resources Institute and World Business Council for Sustainable Development, Geneva, Switzerland, 2011b.

Yin, R.K.: *Case study research design and methods*. Sage Publications, London, UK, 2009.

Zhu, T.T., Zhang, J. & Liu, D.: A study on the evaluation system of green ICT products. Advanced Materials Research 403-408 (2012), pp. 4061–4067.

CHAPTER 8

Micro-training to support sustainable innovations in organizations

Mariette Overschie, Heide Lukosch, Karel Mulder & Pieter de Vries

8.1 INTRODUCTION

Reduction of energy use plays an important role in sustainability approaches. To achieve innovations for energy use in ICT expertise needs to be shared, knowledge has to be increased, and fitting solutions have to be found and implemented. The capability of the organization to learn from the many search and adaptation processes is decisive for the success of innovation (Rosenberg, 1982). Formal disciplinary training programs, important to maintain and improve standardized procedures, are less suitable to connecting learning in innovation trajectories to everyday tasks within the organization.

We introduce micro-training that showed to be a suitable method to support sustainable innovation processes in organizations (de Vries and Brall, 2008; de Vries *et al.*, 2009; Lukosch *et al.*, 2010; Overschie *et al.*, 2007; 2008a; 2008b). In Section 8.2 we will describe impediments in (sustainable) innovation processes, and further elaborate on the question why learning helps overcome these impediments. In Section 8.3, we will address the challenge for sustainable innovations. The micro-training method will be illustrated in Section 8.4, followed by examples of application in ICT production and service companies. The possibilities and impossibilities of using micro-training in organizations provides insight into the applicability and relative successfulness of the method and will be summarized in Section 8.5.

8.2 LEARNING IN SUSTAINABLE INNOVATIONS

How to deal with innovations for sustainability is an emerging issue for many organizations. Innovation aims at fulfilling needs more efficiently and effectively. It is a process in which new configurations of people, organization, and technologies are forged, which requires mutual adaptation (Callon and Latour, 1981). Despite a large amount of experiences from various (pilot) projects, tools, and methods that have been developed to support innovations in practice, it often appears to be complicated for organizations to implement innovations that have been proven elsewhere. Organizations tend to focus on the similarities of the hardware, but as the social and organizational conditions under which these technologies function are different, proven technologies might still fail as is often experienced in technology transfer (Kebede and Mulder, 2008).

Innovations, therefore, need to be tailored to the specific characteristics of an organization (Rosenberg, 1982). The roles, responsibilities, and relationships with technologies have to be redefined (Callon *et al.*, 1986; Latour, 1987; Lundvall, 1992). Successful implementation of a new technology generally requires a complicated series of technical and organizational adaptations (Lundvall, 1992; Schot, 1992) and processes of trial and error (Dosi, 1982). Organizations should take risks; creativity and groundbreaking ideas cannot be ordered by hierarchic leadership. Innovation asks for active participation of various members of the organization, but "mobilizing high levels of participation in the innovation process is unfamiliar and, for many organizations, relatively untested and apparently risky" (Bessant and Caffyn, 1997, p. 10). Japanese companies, regarded as being masters in continuous innovation, continually turn to their suppliers, customers, distributors, government agencies, and even competitors for any new insights or clues that might contribute to their performance (Nonaka and Takeuchi, 1995). As it is especially the "informal"

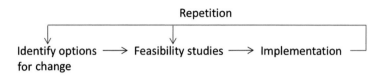

Figure 8.1.   Phases in an innovation process (based on Dieleman and de Hoo, 1993).

organization that is driving innovation, learning should not be limited to the "formal innovators" in an organization (Nelson and Winter, 1982) and that is what we focus on in this chapter. New routines and new heuristics of search processes only gradually establish themselves in the organization (if they are accepted at all). Rogers (1995) described the need for change agents in the innovation-diffusion process. Relevant for repetition of the process is that an innovation should be accepted by a certain minimum number of users in order to be able to conquer the whole market (Rogers, 1995). Continued and targeted information in general is influential in getting sufficient users accepting an innovation (Rogers, 1995).

Attention to learning and change processes is crucial in innovation processes (see Fig. 8.1): it is only through continued application, learning and adoption of feedback that innovations become acceptable alternatives (Rosenberg, 1994). Innovation requires general and specific skills, and explicit and tacit knowledge. Many authors agree that innovation is rooted in people's tacit knowledge and arises through nonplanned bottom-up initiatives as a result of day-to-day experiences (Dosi, 1982; Lundvall, 1992; Nelson and Winter, 1982). Effective transfer of tacit knowledge generally requires extensive personal contact and regular interaction. Nonaka & Takeuchi (1995) argue that for tacit knowledge to be communicated and shared within the organization, it has to be converted into words or numbers that anyone can understand: It is precisely during the time this conversion takes place from tacit to explicit and back again into tacit that organizational knowledge is created. Argyris (1990) explains that if there is no clear need to reflect fundamentally on the tacit assumptions that underlie common patterns of behavior, these are often factored out of the discussion. Constant questioning interferes too much with daily routines. Schön (1993) argues that "when a practitioner does not reflect on his own inquiry, he keeps his intuitive understanding tacit and is inattentive to the limits of his scope of reflective attention . . . . He is unlikely to get very far with his innovation unless he wants to extend and deepen his reflection-in-action, and unless others help him see what he has worked to avoid seeing" (Schön, 1983, pp. 282–283). Again, learning is regarded as an important if not pivotal activity when innovating. Not only learning new things but especially un-learning old routines and reflecting on outcomes seems to be important determinants of successful innovations.

8.3   THE CHALLENGE FOR SUSTAINABLE INNOVATIONS

Sustainable innovations aim for more efficient and effective need fulfillment and diminished resource consumption and pollution, both of products and of the production of these products. Sustainable innovation processes have similar impediments as regular innovation trajectories. Dieleman (1999) studied the gap between the mere existence of opportunities for cleaner production and the realization of these opportunities in projects in The Netherlands. He concluded that in between the two, there is a complex process of acting, searching, learning, and decision making. Overall the projects studied resulted in the establishment of many options for cleaner production, but the implementation of options proved to be rather complicated. In almost all cases, the search processes to tailor the options available, and adapt them to the organization and the existing production processes were not easy because:

- It was not part of people's jobs and standard responsibilities;
- People could not rely on routines and known ways of operating, interaction, and communication;

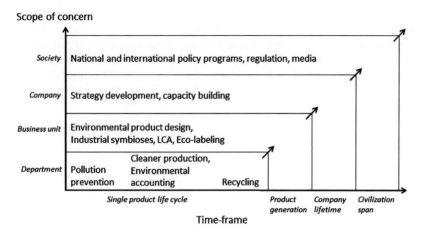

Figure 8.2.   The challenge for sustainable innovation (based on Mulder, 2006).

- Responsible persons showed a lack of conviction in trying to involve employees. They did not set clear objectives and therefore it was not clear why things should be done and whether it was worth committing resources;
- Almost no support and technical assistance could be found from the business environment: the various stakeholders, like consultants, trade organizations, and research organizations (Dieleman, 1999).

Experiences in a program about Corporate Social Responsibly (CSR) in The Netherlands entitled "From financial to sustainable profit" showed that the process of getting support from top to bottom in the organization was one of the hardest tasks in the whole endeavor (Cramer, 2003). A group of 19 Dutch companies (of which three ICT oriented) participated in the discussions organized by the National Initiative for Sustainable Development (NIDO) on issues raised by the company personnel themselves. These were related to knowledge gaps or were concerned with problems encountered in the process of implementing CSR. Moreover, every company carried out its own internal CSR project. "Making people enthusiastic and creating internal backing took up a lot of time" (Cramer, 2003, p. 136). Cramer recommended to further elaborate process-oriented instruments to create support within the organization for CSR and to support measures for implementation of CSR in small and medium sized (SME) companies (Cramer, 2005, p. 592).

Low hanging fruits through the implementation of incremental changes are important and seem to be easy to grasp. More difficult are radical changes with a longer payback time, especially in sectors which account return on investment for only half a year. Far-reaching radical technological changes are perhaps not eco-competitive now, but at a certain moment they will begin to deliver a higher reduction in environmental impact (Cramer, 2000). All these levels of action can be taken by organizations to contribute to the pathway of reaching an eco-leap.

Mulder (2006) frames the challenge for sustainable innovation as a consistent attempt to link various levels of environmental performance, see Figure 8.2. The first environmental improvements in an organization generally focus on manufacturing and maintenance. Methods such as cleaner production and environmental accounting could be complemented by pollution prevention, and recycling, the last taking a longer timeframe. At the level of business unit and product design, these methods could be complemented with environmentally optimized product design and industrial symbiosis, life-cycle assessment (LCA) and eco-labeling. Finally to achieve the required leaps in eco-efficiency, this should be embedded in a long term corporate strategy and national policies. The mere adoption (or improvement) of existing technology via continuous improvements will be insufficient in the long run. We need leaps in eco-efficiency to provide scope for the development of the underdeveloped world without ruining the planet even more (Jansen, 1994) and to create room for more radical eco-effective innovations that can generate value over a long period of time. Strategic decisions in organizations cannot be made by calculations alone as they always involve a choice for the kind of organization and the kind of society we want to create for the future (Jansen, 1994; Mulder, 2006). Long term effects (as well as "long distance effects") are likely to remain unobserved by the acting individual, even if feedback is improved (Leeuwis *et al.*, 2004), the challenge for sustainable innovations is therefore to seek for a more long term perspective that allows not only actors to identify options for reduced resource consumption and reduced pollution but also to involve more stakeholders than are taken into account thus far (Mulder, 2006). Organizations are increasingly confronted with societal issues by employees having a private life and societal issues tend to percolate to the work floor. Customers demand supply of more sustainable products, and neighbors require a clean environment, fear safety issues, etc. The organization should be capable to understand its work in a wider context. To innovate successfully, it should be able to look further than its immediate scope of concern and this requires learning that goes over departmental and organizational boundaries.

8.4   MICRO-TRAINING TO SUPPORT LEARNING IN ORGANIZATIONS

In this Section we illustrate why the micro-training method stimulates learning to overcome barriers in sustainable innovation processes. But before we detail how you might use micro-training in your organization we will describe the didactical models behind this method.

Learning is promoted when:

- learners are engaged in solving real-world problems,
- knowledge is activated as a foundation for new knowledge,
- new knowledge is demonstrated to the learner,
- new knowledge is applied by the learner, and
- new knowledge is integrated into the learner's world (Merrill, 2002).

Kirschner *et al.* (2006) contrasted guided models, such as direct instruction with minimally guided methods, such as discovery learning, problem-based learning, inquiry learning, experiential learning, and constructivist learning. They showed that providing only minimal guidance during instruction "does not work" and results are much better if a teacher takes the role of activator instead of facilitator. Direct instruction guidance is providing information that fully explains the concepts and procedures students are required to learn, as well as the learning strategy (Kirschner *et al.*, 2006). Gagné *et al.* (1992) identified five types of learning outcomes of direct instruction: verbal information, intellectual skills, cognitive strategies, motor skills, and attitudes. Adams and Engelmann (1996) view direct instruction as a suitable learning method to accelerate learning. Although Gagné and Kirschner mainly focus their research on formal education for predominantly young learners, their work provides insight that helps select an effective training method. Effective training or instruction in the organization should be about topics that make sense to the participants (appropriateness), are of interest to them (motivation), and in a way that the topic is easy to remember and apply (good quality of training) (Slavin, 1996).

In case of innovation processes it is often not clear what has to be learnt. To support flexibility and sharing of both tacit and explicit knowledge at the point of need we developed a method to organize short training sessions (Overschie, 2007). In the development and testing of the training method, organizations of various scales, representing a variety of technology oriented activities and core businesses, have cooperated. The organizations desired to involve all kind of target groups in sustainable innovations. Requirements for the development of the method were:

- Training is facilitated by employees who are not trainer by profession;
- The training-topic is defined by the employees involved;
- Training takes place on or near the workplace;
- Training is organized during appropriate moments;
- The training method is usable for various target groups.

In medium-sized and small scale groups micro-training fosters an active learning process with a minimum of interruptions to the normal workflow. The aim is not merely focusing on individual learning but to support social processes and change. Micro-training is a framework for types of self-directed learning support with a high level of practical relevance to be used by the employee or manager. Knowles (1975) defined self-directed learning as "a process in which individuals take the initiative, with or without the help of others, in diagnosing their learning needs, formulating learning goals, identifying human and material resources for learning, choosing and implementing appropriate learning strategies and evaluating learning outcomes" (Knowles, 1975, p. 18). The didactical grounds of direct instruction are used to establish effective training, facilitated by employees who are not training experts. The micro-training topics can be chosen by the learners involved. The short duration of the session offers flexibility in planning, which links to the principle of learning at the point of need, or so-called teachable moments. The issue of the necessity of time for knowledge acquisition is addressed by dividing the topic of training into sub-topics, linked in series of short sessions.

A session comprises a time span of about 15 min, which can activate and maintain learning processes for a longer period if they are bundled up in series, being face-to-face, online, or in an e-learning situation. Because of the richness of the human space, contact and interaction in face-to face meetings are superior to online meetings (Siemens, 2006).

The sessions can be organized quickly by an employee, the so-called micro-trainer. We take a constructivistic approach (Jonassen, 1991) in the design of the training sessions with a focus on the learning needs of the participants. Each session will be structured in the same way. It starts active, for example by addressing an actual fact or raising a question, followed by an exercise or demonstration, and discussion including feedback. It ends with directions for further development and a brief preview of the next sessions (see Fig. 8.3).

- The aim of an *active start* is to really involve the participants. This start should consist of a "mental activity" such as reflecting and making up your own mind.
- The *exercise* contributes to sharing knowledge. This part contains the information that needs to be dealt with.
- During the *discussion part,* feedback will be provided and there will be checked if all participants really understand the topic of the session and if there are questions.
- The last part "*How to continue*" aims at stimulating involvement and ensures that participants leave with a clear goal. Follow up on the (sub)-topic is needed.

It is important to realize that, because of its decentralized ad-hoc nature and execution by relatively autonomous working units that integrate the sessions and its outcomes in the daily

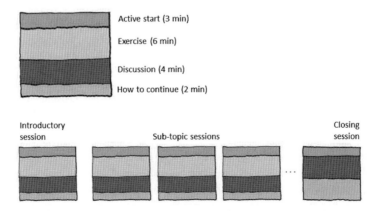

Figure 8.3.   Series of micro-training sessions (based on Overschie *et al.*, 2007).

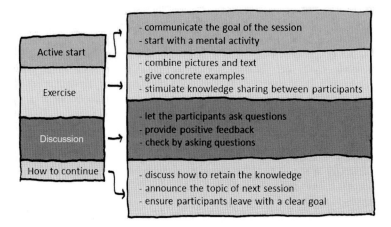

Figure 8.4.   Process worksheet of a micro-training session (based on Overschie *et al.*, 2007).

workflow, micro-training cannot be externally supervised and controlled. Two cases will show how micro-training positively involves employee's knowledge in innovation processes through mobilization of employees.

8.4.1   *Case 1: Micro-training in a mechatronics company*

To remain successful, an innovative SME, developer and manufacturer of mechatronics, part of a large international holding, spends a lot of time on training and educating its staff. However, there are issues that deserve additional attention such as increasing the knowledge of production workers, sharing production knowledge that is relevant to the R&D department, and training new employees.

All employees were informed on micro-training during a presentation of the managing director. Afterwards a continuous slide show was displayed in the entrance hall. At the start the director acted as a so-called micro-trainer. He prepared the sessions and invited three to four production workers and a colleague from R&D. The sessions aimed to reduce product failures and increase productivity. They were planned ad-hoc, depending on the availability of time slots in his agenda. The topics were based on real needs and were organized a number of times to offer all involved employees the opportunity to participate (see Fig. 8.5, left side). Brainstorming and discussion techniques were used to stimulate new ideas. The managing director used slides with visuals to focus the sessions. He raised questions and encouraged all participants to join the discussion. A flip-chart was used to support the dialogue and to make an inventory of the ideas. The managing director introduced the didactics of micro-training to different colleagues so they could also apply these in an afternoon workshop with all colleagues, on the topic of quality improvement. They used the results of the different sessions as input for the next session (see Fig. 8.5, middle). In several sessions, small groups of four to five people from various departments (work floor, R&D and Management) discussed the quality of various products, the possibility for improvements, and their contribution to the realization of the options. The group results were presented and the overall results have been transformed into an action plan. After this workshop different employees could organize sessions at random and "spontaneously" to solve existing issues. They invited colleagues that could contribute to the solution or might need to learn on specific issues (see Fig. 8.5, right side).

8.4.2   *Case 2: Micro-training in a service company for computers*

The sales manager of a small-size service company for computers in Sweden was looking for a method to better share information in his daily work and when working in projects. After a one

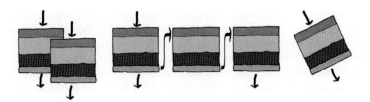

Figure 8.5.   Micro-training in a mechatronics company.

hour training facilitated by a micro-training expert, he was able to use the micro-training method in his own organization. The sales manager had regular sales meetings at which micro-training has been applied. The method was launched just as a change of the meeting structure. There was no documentation; he only opened up with a question concerning the chosen topic and let the participants answer it. In some cases, somebody else was given the role to shortly talk about the subject. The participants got very active during the sessions. Since these were ordinary parts of the sales meetings, the micro-training could easily continue covering specific topics that the sales manager and his crew wanted to discuss.

One of these topics was the implementation of a "green concept". A computer monitor made of 65% recycled plastic is part of this concept. The company collaborated with the monitor producer bringing the new monitor to the market. A series of sessions has been created around the question: What does the green concept mean to us? Sub-topics were about: What does the green monitor imply for us, for the customer and for the producing company? or Energy consumption of computers—how can we affect this? As an active start the sales manager addressed his view on responsibility and follow-up, followed by a short discussion about this view, ending in a question to the participants: "What is your idea?" The participants were invited to think of this until the next session when one of them presented his/her view. Furthermore, the environmental expert of the service company used micro-training in order to supply the salesmen with environmental arguments towards the customers.

The short time of 15 min and the focus on only one issue are seen by the sales manager as the main advantages of using micro-training. The sales manager also drew the conclusions that it was important to be consistent when applying the method, and that the session should be well prepared. Topics are almost inexhaustible since there are always new things to focus on. Micro-training is short, easy, and focused. According to the sales manager it would be good if the method could be standardized in the organization and used as a tool.

8.5   CONCLUSIONS AND REFLECTION

With regard to sustainable innovations organizations should learn to take a wider range of stakeholders into account. Micro-training is a mean to involve employees in this process and to create social interaction to support change. The method aims to mobilize participation, to bridge the gap between employees with various responsibilities in the organization, and to stimulate cross-departmental collaboration. The cases show the potential micro-training holds to share knowledge and transfer learning experiences both in SME's and departments of large organizations. Communication occurs during concrete social interaction embedded in the context of the organization, which supports the integration of both explicit and tacit knowledge.

The added value of micro-training is the support of a "*can-do*" mentality where sustainability issues in the organization arise. The sessions can be carried out by any person in the organization that can provide the expertise or skills and offers flexibility in the choice of topics and planning but has a clear, easy-to-learn process structure. Micro-training leads initially to incremental improvements at the departmental level and has the potential to evolve to broader changes that ask for collaboration between various departments. Learning from exchanging knowledge and experiences, and acquiring and applying new knowledge are a vital condition in this respect. The short series of sessions offer the opportunity to address topics at the point of need. Keeping a topic on the agenda helps widening the employees' perspective, and makes employees see the advantages of alternative approaches, not just the risks. Short sessions will never replace formal training courses that provide the employees with large chunks of conceptual knowledge, but micro-training can help support learning for sustainable innovations since over time time incremental improvements focusing on efficiency remain crucial for the more fundamental eco-effective innovations.

ACKNOWLEDGEMENTS

Thanks to Arno van Wayenburg, Ton van der Voort van der Kleij, Lennart Sundberg, and Monika Olsson for their contribution to the cases.

BIBLIOGRAPHY

Adams, G.L. & Engelmann, S.: *Research on direct instruction: 25 years beyond DISTAR*. Educational Achievement Systems, Seattle, WA, 1996.

Argyris, C.: *Overcoming organizational defenses*. Prentice Hall, New Jersey, 1990.

Bessant, J. & Caffyn, S.: High-involvement innovation through continuous improvement. *International Journal of Technology Management* 14:1 (1997), pp. 7–28.

Callon, M. & Latour, B.: Unscrewing the big leviathan: how actors macro-structure reality and how sociologists help them to do so. *Advances in social theory and methodology: toward an integration of micro-and macro-sociologies*, 1981, pp. 277–303.

Callon, M., Law, J. & Rip, A.: *Mapping the dynamics of science and technology*. Springer, Berlin, Heidelberg, 1986.

Cramer, J.: Responsiveness of industry to eco-efficiency improvements in the product chain: the case of akzo nobel. *Business Strategy and the Environment* 9:1 (2000), pp. 36–48.

Cramer, J.: *Learning about corporate social responsibility*. IOS Press, Amsterdam, The Netherlands, 2003.

Cramer, J.: Experiences with structuring corporate social responsibility in Dutch industry. *Journal of Cleaner Production* 13:6 (2005), pp. 583–592.

de Vries, P. & Brall, S.: Microtraining as a support mechanism for informal learning. *eLearning Papers* 11:1, Barcelona, Spain, (2008).

de Vries, P., Lukosch, H. & Overschie, M.: Microtraining as an effective way towards sustainability. *Proceedings of Edulearn09 International Association of Technology, Education, Development*, 2009, pp. 31–38.

Dieleman, H.: De arena van schonere productie, mens en organisatie tussen behoud en verandering (*The arena of cleaner production, man and organization between conservation and change*.) PhD Thesis, (in Dutch with English summary), Eburon, Delft, The Netherlands, 1999.

Dieleman, H. & de Hoo, S.: Toward a tailor-made process of pollution prevention and cleaner production: results and implications of the prisma project. In: K. Fisher & J. Schot (eds): *Environmental strategies for industry*. Island Press, Washington, DC, 1993, pp. 245–275.

Dosi, G.: Technological paradigms and technological trajectories: a suggested interpretation of the determinants and directions of technical change. *Research Policy* 11:3 (1982), pp. 147–162.

Gagne, R.M., Briggs, L.J. & Wager, W.W.: *Principles of instructional design*. 4th ed, Hartcourt, Brace Jovanovich, Publishers, Orlando, FL, 1992.

Jansen, L.: Towards a sustainable future, en route with technology! In: *The Environment: towards a sustainable future*. Springer, 1994, pp. 496–525.

Jonassen, D.H.: Objectivism versus constructivism: do we need a new philosophical paradigm? *Educational Technology Research and Development* 39:3 (1991), pp. 5–14.

Kebede, K.Y. & Mulder, K.: Needs assessment and technology assessment: crucial steps in technology transfer to developing countries. *Revista Internacional de Sostenibilidad, Tecnología y Humanismo* 3 (2008), pp. 85–104.

Kirschner, P.A., Sweller, J. & Clark, R.E.: Why minimal guidance during instruction does not work: an analysis of the failure of constructivist, discovery, problem-based, experiential, and inquiry-based teaching. *Educational Psychologist* 41:2 (2006), pp. 75–86.

Knowles, M.S.: *Self-directed learning – A guide for learners and teachers*. Associated Press, New York, 1975.

Latour, B.: *Science in action: How to follow scientists and engineers through society*. Harvard University Press, Cambridge, 1987.

Leeuwis, C., & van den Ban, A.: *Communication for rural innovation*. Blackwell Science Ltd, Oxford, UK, 2004.

Lukosch, H., Overschie, M., Olsson, M. & de Vries, P.: Supporting sustainable development through micro-training. *Proceedings of the International Conference Engineering Education in Sustainable Development (EESD 2010)*, 2010.

Lundvall, B. (ed.): *National systems of innovation; towards a theory of innovation and interactive learning*. Pinter Publishers, London, UK, 1992.

Merrill, M.D.: First principles of instruction. *Educational Technology Research and Development*, 50:3, Springer, US, (2002), pp. 43–59.

Mulder, K.: *Sustainable development for engineers: a handbook and resource guide*. Greenleaf Publishing, Sheffield, UK, 2006.

Nelson, R.R. & Winter, S.G.: *An evolutionary theory of economic change*. Harvard Business School Press, Cambridge, 1982.

Nonaka, I. & Takeuchi, H.: *The knowledge-creating company: how Japanese companies create the dynamics of innovation*. Oxford University Press, USA, 1995.

Overschie, M.: Microteaching manual: effective transfer of knowledge for sustainable technological innovation. 2007, http://www.microteaching.org (accessed January 2013).

Overschie, M., de Vriesb, P., Pujadasc, M. & van Wayenburg, A.: Microteaching for sustainable technological innovation in companies user experiences from The Netherlands and Spain. *Proceedings of the 11th European Roundtable on Sustainable Consumption and Production conference (ERSCP 2007)*, Basel, Switzerland, June 20–22, 2007.

Overschie, M., Pujadas, M., Espuna, A. & de Vries, P.: Microteaching to support incremental innovations for sustainability. *Greening of Industry Network 2008*, Leeuwarden, Netherlands, 2008a.

Overschie, M., de Vries, P., Espuna, A. & Weingarts-Gottgens, K.: Effective learning to support incremental innovations towards sustainability. In: Braunegg S. (ed.): *Bridging the gap*. Engineering Education in Sustainable Development 2008 Conference, 22nd to 24th September, Graz University of Technology, Graz, Austria, 2008b, pp. 520–528.

Rogers, E.: *Diffusion of innovation*. 4th ed, The Free Press, New York, 1995.

Rosenberg, N.: *Inside the black box: technology and economics*. Cambridge University Press, Cambridge UK, 1982.

Rosenberg, N.: *Exploring the black box: technology, economics, and history*. Cambridge University Press, Cambridge UK, 1994.

Schön, D.A.: *The reflective practitioner. How professionals think in action* (Vol 5126). Basic Books, New York, 1983.

Schot, J.W.: Constructive technology assessment and technology dynamics: the case of clean technologies. *Science, Technology & Human Values* 17:1, Sage Publication, London, UK, (1992), pp. 36–56.

Siemens, G.: Knowing knowledge, Lulu Press, Morrisville, NC, 2006.

Slavin, R.E.: Education for all. *Contexts of Learning Series*. Educational Achievement Systems. Seattle, WA, 1996.

CHAPTER 9

Integration of green IT knowledge in education

Henk Plessius

9.1 INTRODUCTION

The increasing interest in sustainability in the world of IT has not yet fully found its way to education. Contrary to sectors like engineering and chemistry, where sustainability seems incorporated quite generally in the curricula nowadays, it hardly is a topic in the IT curricula of higher education. A quick scan conducted by the author in The Netherlands in January 2012 (Plessius, 2012), showed that in the higher educational sector only in 10% of all IT curricula sustainability had gained a systematic (and measurable) place. This may well become a bottleneck for the greening of the IT sector, as academic and professional research are necessary conditions for innovation and without innovation, sustainability and reduction of energy-use will stall.

This apparent lack of interest towards sustainability in IT education may be due to the technological rat race in the sector: hardware and software technologies quite often are outdated within a couple of years, making it a major task to offer a more or less up-to-date curriculum to students. There are however many reasons why sustainability should be taken into account in the IT curricula, the two foremost being energy consumption and electronic waste (e-waste or e-scrap). There is however a strategic reason as well: sustainability becomes increasingly important for business continuity. Let us look at some examples that emphasize the importance of green IT solutions.

> As we mark the fifth anniversary of our annual study of the digital universe, it behooves us to take stock of what we have learned about it over the years. We always knew it was big—in 2010 cracking the zettabyte barrier. In 2011, the amount of information created and replicated will surpass 1.8 zettabytes (1.8 trillion gigabytes)—growing by a factor of 9 in just five years.
>
> (Source: Gantz, J. and Reinsel, S. Extracting Value from Chaos. IDC iView. June 2011. See http://www.emc.com/digital_universe (Gantz and Reinsel, 2011)).

As revealed in a study by IDC (Gantz and Reinsel, 2011), the amount of stored information has grown with almost 70% each year for the last five years and it is expected to grow by at least the same percentage in the next decade. This growth rate implies a staggering 100-fold increase of digital data for the next decade—almost beyond belief if we had not realized a similar growth in the last 10 years! This growth has been made possible by a yearly price-reduction for storage of 50% and exploding budgets for IT in enterprises.

Without technological advancements, the power consumption of the equipment needed for storing and handling of all these data would have increased a 100-fold as well. Luckily, technology did not stagnate: a survey carried out in 2007 at Stanford University (Koomey, 2007) shows a relatively moderate twofold increase in power consumption over a 5-year period. Extrapolating, this means that in 10 years time, the energy consumption of the average server park will have to increase by a factor 4 to handle the predicted data growth. We may be able to cope with that growth for the decade to come, but it is obvious that yet another 10 years of such growth in energy consumption is not feasible. Therefore, the challenge is to meet the increasing demand for IT, while on the other hand staying within the current power envelope (see, for instance, the Chapter 7 written by Kalsheim and Beulen in this book). This means that "power consumption" will become a constraint like "performance", "privacy", "responsiveness", etc. and we need a new generation of IT-specialists who know how to deal with this phenomenon.

The second reason to incorporate sustainability in IT curricula concerns e-waste: electronics used in computers and other IT-components contain a large amount of rare (and quite often poisonous) materials like gold, copper, cadmium, etc.

> *The category of "selected consumer electronic products" grew by almost 5% from 2007 to 2008, from 3.01 million tons to 3.16 million tons. While it's not a large part of the waste stream, e-waste shows a higher growth rate than any other category of municipal waste in the EPA's report. Overall, between 2007 and 2008, total volumes of municipal waste DECREASED, while e-waste volumes continue to increase.*
>
> (Source: Electronics Takeback Coalition. Facts and Figures on E-waste and recycling. Jan 2011. See www.electronicstakeback.com (Coalition, 2011)).

We cannot continue to waste these scarce materials as we do today thereby endangering human life and precious land (recycling—if any—is mainly done in Africa and Asia because of the more lenient legislation in those regions). We can change our ways and reuse, refurbish, or recycle, rather than just dispose of these electronic components. Education could contribute to effectuate such changes.

Where energy consumption and e-waste are sustainability aspects intrinsic to present-day IT, they should be overcome within the sector itself: the greening of IT. However, IT is increasingly used for greening in other sectors as well (known as greening by IT) as the following text shows:

> *Tomlinson describes many efforts toward sustainability supported by IT—from fishers in India who maximized the sales potential of their catch by coordinating their activities with mobile phones to the installation of smart meters that optimize electricity use in California households and offers three detailed studies of specific research projects that he and his colleagues have undertaken: EcoRaft, an interactive museum exhibit to help children learn principles of restoration ecology; Trackulous, a set of web-based tools with which people can chart their own environmental behavior; and GreenScanner, an online system that provides access to environmental-impact reports about consumer products. Taken together, these examples illustrate the significant environmental benefits that innovations in information technology can enable.*
>
> (Source: Tomlinson, B. Greening through IT. Information Technology for Environmental Sustainability. MIT Press, Cambridge, Massachussetts. 2010 (Tomlinson, 2010)).

In cases like the one above, the disadvantages of more IT in terms of sustainability are well compensated for by the application. Therefore, it can be argued that from a sustainability viewpoint more IT sometimes is better, particularly in cases where the use of IT increases overall sustainability.

Finally, responsibility towards society is a driver towards more sustainability in business. Robert Pojasek has expressed this in 2010 as follows:

> *Every organization depends on maintaining a license to operate from society. This does not refer to any specific document or permit required by law. Rather, it refers to society's acceptance of the organization and its willingness to let the business continue operating. Companies gain (and Maintain) their licenses to operate by providing value to the community. In this context, value refers to more than just economic development. It also encompasses environmental protection and social equity. These three concerns (environment, society and economics) represent the three areas of responsibility that increasingly are being assigned to business.*
>
> (Source: Pojasek, R.B. Sustainability: The three responsibilities. Environmental Quality Management, 19: 87–94. 2010 (Pojasek, 2010)).

In this introduction, we have shown various reasons that support the greening of IT, ranging from the reduction of energy and material waste to business continuity, the "license to operate" of the sector. For these reasons, we propose that sustainability should be part of every IT curriculum. How to incorporate it and what is relevant from an educational perspective, are questions we will address in this chapter. To that end, we first focus on green IT topics and ask ourselves, what are relevant sustainability issues for educational programs in IT from a technological point of view. While doing so, we will introduce the *green IT framework*. Second, we look at sustainability from an educational viewpoint: what are the major sustainability competencies that should be embedded in an IT curriculum and how do these relate to the IT lifecycle? In the third and final part of this chapter, we will combine these two viewpoints and propose a "sustainability-proof" IT curriculum, giving examples how green IT issues can be integrated in the regular IT curricula.

9.2   THE GREEN IT FRAMEWORK

To establish which aspects of sustainability are relevant in an IT curriculum, we will look in more detail towards the field of green IT in order to categorize these aspects. In general, the term green IT refers to environmentally sustainable computing. In an overview of the field, San Murugesan defines green IT more precisely as: "*the study and practice of designing, manufacturing, using, and disposing of computers, servers, and associated subsystems such as monitors, printers, storage devices, and networking and communications systems efficiently and effectively with minimal or no impact on the environment. Thus, green IT includes the dimensions of environmental sustainability, the economics of energy efficiency, and the total cost of ownership, which includes the cost of disposal and recycling*" (Murugesan, 2008).

This definition refers to the greening of IT, but does not include the greening by IT. For our purpose, we will broaden it to incorporate the greening by IT as well. In our view, green IT is concerned with minimizing the environmental effects of products and services that include IT components over the course of their lifecycle. In order to assess the environmental effects of these products and services we will look in more detail at IT components and the product lifecycle first.

IT components are the constituents of an IT system and as such may consist of any combination of hardware, software, data, and knowledge. Such a component may be categorized in one of the following three fields:

1. Infrastructure: the shared components, not to be touched by users. These components are typically found in data and computing centers and may be enterprise owned or cloud based. In most cases, these components are referred to by their functionality: data storage, software services, secure access, web hosting, etc.
2. Local: IT for the end-user. Typical hardware examples are desktop computers, laptops and tablets, smart telephones, printers. This category includes end-user applications like spreadsheets and word processing as well as apps and locally stored information.
3. Transport: the components used for the exchange of data between various nodes of a network, local and/or infrastructure. Examples are not only hardware like routers, switches, and cabling, but drivers and protocols like IPv6 as well. These components may be seen as part of the infrastructure as well, but, complying with the current practice in IT, we prefer to set them apart.

Components may be found in a wide range of different products and services. We follow (Murugesan, 2008) and divide the lifecycle of these products and services in four phases:

1. Design: the initial phase of a product: from idea to blueprint.
2. Manufacturing: where the actual product is realized.
3. Use: the period in which the product is actually in use.
4. Disposal: the end of a product's lifecycle; the product may be refurbished, recycled, or just disposed of.

In every phase of a product's lifecycle, IT components from each of the three fields may be present as each of these components has to be designed, manufactured, put to use, and be disposed

Phase	Field	Infrastructure	Transport	Local
Design				
Manufacturing				
Use				
Disposal				

Figure 9.1.   The green IT framework.

of. Therefore, we can represent the field of green IT in a 3 by 4 matrix (see Fig. 9.1): the green IT framework.

In every cell of the matrix, we assess sustainability aspects of the components with regard to all relevant aspects in terms of energy and material usage, making it feasible to review its relevance for inclusion in an IT curriculum.

Having approached sustainability in IT by combining a lifecycle point of view with a threefold subdivision of the total IT system, we will now elaborate on the educational angle, thereby introducing educational terminology relevant to our topic. By doing so, we aim to give policymakers and educators in business and education practical tips on how to "green" workshops, courses, and curricula.

9.3   COMPETENCIES IN GREEN IT

Nowadays, it is increasingly customary to describe learning outcomes of a curriculum as competencies to be acquired. A competency is normally associated with a combination of knowledge, skills, and attitudes appropriate in a given context, defined by Dochy & Nickmans (Dochy and Nickmans, 2005) as: "*a competency is a personal capability that becomes visible by showing successful behavior in a specific context*". A competency thus integrates three elements: knowledge, skills, and attitude and may combine technical knowledge with general problem solving skills and social-communicative behavior. Competencies are usually derived by analyzing the typical products delivered by professionals in the line of their work and such competencies should adhere to professional standards. When a competency is applicable in a broader domain (e.g. "uses a problem-oriented approach"), it is called *domain-general*. If on the other hand, a competency is specific for a certain profession (e.g. "able to program in C++ on the highest level") it is called *profession specific*.

Depending on the line of work and the complexity of the job, a competency may be practiced at different proficiency levels. In this chapter, we will use a scale of 0 to 4[1] meaning:

0 Awareness: aware of the basic facts, but not capable of applying these in practical situations.
1 Basic: able to apply knowledge and skills to solve straightforward problems.

---

[1]The levels 1 to 4 correspond with the levels e-1 to e-4 of the European e-Competence Framework (EEC, 2010). We have added a level 0 (awareness) as this is the first level to be reached in any topic and as such an important step in education.

2 Advanced: uses knowledge and skills to solve problems within a predictable context.
3 Professional: uses knowledge and skills to solve complex problems within a (sometimes unpredictable) context.
4 Expert: can transfer knowledge and skills to solve complex problems within a new context.

A good example of a set of competencies for the IT sector is the European e-Competence Framework (EeCF) version 2.0 (EEC, 2010) that discerns 36 profession-specific competencies built around five areas in the IT business process: plan, build, run, enable, and manage. In this framework, we find one competency dealing with sustainability:

A.8. Sustainable Development

Estimates the impact of ICT solutions in terms of eco responsibilities including energy consumption. Advises business and ICT stakeholders on sustainable alternatives that are consistent with the business strategy. Applies an ICT purchasing and sales policy which fulfils eco-responsibilities.

(Source: European e-Competence Framework version 2.0, September 2010. See http://www.ecompetences.eu/ (EEC, 2010)).

This competency has been categorized in the plan area of the EeCF and it is advised to strive for a proficiency level of 3 to 4.

While we welcome the explicit recognition of the importance of sustainability for IT, we feel that sustainability issues are relevant in the other areas of the EeCF (build, run, enable, and manage) as well. However, in current competency sets for IT, the topic is not included at all or only in very general terms that are not IT-specific. A concise overview of sustainability competencies (not restricted to IT only) and related issues is given by Willard *et al.* (2010) who have interviewed professionals in industry on the top skills needed for success as a sustainability professional. From these interviews, they learned that strategic planning, systems thinking, and project management were considered as the most important hard skills while communication with stakeholders, problem solving, inspiring, and motivating others were considered as the most important soft skills. These are important domain-general competencies and as such should be included in the sustainability competencies for IT curricula as well. How to do this will be the subject of Section 9.4.

9.4   GREENING OF IT CURRICULA

When bringing green IT elements into higher education IT curricula, it is important to discern between profession-specific competencies and more domain-general competencies. Domain-general competencies are important for every IT professional but their proficiency level is generally low: awareness (level 0) or sometimes basic (level 1) on the scale given above. Profession-specific competencies on the other hand are dependent on the line of work and usually a proficiency level of advanced (2) to professional (3) is aimed for, depending on the type of job and the ambition of the educational institute. These competencies (and their desired proficiency level) can only be derived from a careful analysis of the domain of the study as they are directly related to that domain.

Let us look at the domain-general green IT competencies first. IT professionals—as most knowledge workers—are dependent on IT in the routine of their job. Contrary to other professionals, a certain understanding of how these systems work, is expected from IT professionals. In education, this knowledge can easily be extended to include sustainability issues as well, thereby creating an awareness of energy consumption and material usage that may be put to use in daily behavior. Moreover, as IT students acquire a basic understanding of data processing and data

Field / Phase	Infrastructure	Transport	Local
Design			– Is looking for green IT solutions (0)
Manufacturing			
Use	– Is aware of energy consumption (0) – Is alert on greening by IT (0)		– Reduces energy consumption (1) – Is alert on Greening by IT (0)
Disposal			– Reduces e-waste (1)

Figure 9.2.   Domain-general green IT competencies.

transport (internet), areas where nowadays energy consumption plays a very important role, techniques for the reduction of power consumption can be integrated. Last but not least, it is important that every IT student learns to be alert to the possibilities IT offers for greening other sectors. In Figure 9.2, we have depicted some domain-general competencies and their proficiency levels in the green IT framework.

We put forward that every student should at least comply with the following green IT competencies (between the brackets, the desired proficiency levels are given):

IT1 – Awareness of the energy consumption of IT in daily use by hardware, software, and applications (0).

IT2 – Ability to measure energy consumption of local IT components and, where possible, to reduce this and/or to choose a low-energy alternative (1).

IT3 – Familiar with the constituents of IT components and biased towards the most sustainable components to reduce e-waste (1).

IT4 – Alertness on the possibilities of IT to gain in general sustainability and biased towards green IT solutions (0).

To introduce these general competencies, at our university a virtual world has been built where undergraduate students will be given assignments[2] regarding sustainability. The photograph in Figure 9.3 shows one of the buildings in that world (which can be found on OpenSim (www.opensimulator.org), grid Virtyou, World GreenIT).

On the other hand, profession-specific green IT competencies are more complex to deal with as they are connected closely to the competencies relevant in the specific domain in which the student is trained. For example, a network engineer should learn more about the transport field whereas for a software engineer the design and manufacturing phases are most important and for an information manager the business aspects in the Local field may be of prime importance.

---

[2]For example: computers that are not in use, should be turned off to save on energy.

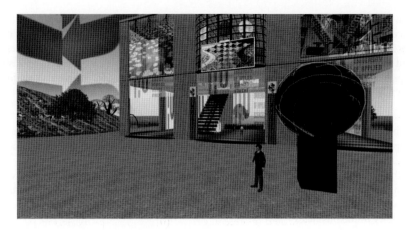

Figure 9.3.   The GreenIT world on OpenSim.

To give an impression of how green IT competencies may be incorporated in these curricula and thus contribute to effective and efficient solutions, we will give some examples from various IT domains.

9.5   EXAMPLES OF GREEN IT IN EDUCATION

We concentrated on green aspects of IT relating to content. This does by no means imply that new courses should be developed in an existing curriculum. On the contrary, we think it is important to integrate the green IT competencies in existing courses. For example, an institution may offer a beginner's course on Information Technology. The topic of power consumption may be added when various devices are discussed. In this way, "green" becomes a logical extension of IT. In the same way, questions involving energy and/or material usage may be added in cases and projects. Moreover, whenever a project involves another domain, aspects of greening by IT can be emphasized. In this section, we will give some examples in various domains showing how this can be done. All examples are from courses at our university[3], where sustainability is a key issue in operational management as well as education.

9.5.1   *Measuring and reducing power consumption*

For part-time bachelor students Systems and Network Engineering (the students are working as System Administrators for four days a week and attend classes the fifth day), a course on data center technologies is part of the curriculum. The focus of the existing—technology-driven—course is on virtualization and data storage techniques. We decided to extend this course with green IT aspects, in particular energy consumption, as this is a major topic in data centers nowadays[4]. Goal is to extend the competency IT2 (Ability to measure energy consumption of local IT components and, where possible, to reduce this and/or to choose for a low-energy alternative) towards infrastructural components and at the same time increase the proficiency level of the students to level 2 (advanced). In our green IT framework, this can be mapped on the yse phase in the Infrastructure field.

---

[3]The University of Applied Sciences Hogeschool Utrecht, see http://international.hu.nl
[4]For example, the Green Computing Initiative (See http://www.greenci.org) offers various training programs and certificates for IT professionals in systems and networking.

Given the specific situation of the students (employees in different companies), we decided to introduce an assignment in their own job environment. As these environments were quite different, students had to start with a proposal including their research question. Common concepts to be included in these proposals were data storage and virtualization, measuring energy consumption and steps to reduce energy consumption. Stated differently, the student proposals should comply with the following learning outcomes:

- Has thorough knowledge of data storage techniques (specifically SAN/NAS) that shows in the ability to generate benefits and drawbacks in specific situations, when these techniques are applicable and what their contribution towards sustainability is, inside the data center and concerning continuity of the business.
- Has thorough knowledge of virtualization techniques that shows in the ability to give mutual dependencies with respect to data storage techniques, when these techniques are applicable and what their contribution towards sustainability is inside the data center and concerning continuity of the business.

Once the proposals are approved, the actual research can start and the results should be presented in a paper.

At the start of the assignment, students had their reservations; typical arguments were "in our company all necessary arrangements are already made to reduce power consumption" and "business continuity is priority number one and cannot be combined with sustainability". However, in the end all students were able to propose measures that could lead to a reduction in energy consumption of at least 20%, which changed their view on the topic drastically. The final papers showed that the students had reached the desired proficiency level for the competency as well.

This example shows how education in IT and energy reduction may be joined and lead to benefits for the organizations involved, its employees and students, and society.

9.5.2   *Green IT in information management*

For a second example, we look at the topic of information management. At our university, students can choose information management as their major. As information managers typically are decision makers on IT in organizations, we think it is important to include sustainability issues in their education. However, it is not easy to outline the relevant green IT aspects that are important for every information manager. Information management is "the collection and management of information from one or more sources and the distribution of that information to one or more audiences". As such, information managers may have to deal with various aspects of information that may be summed up by: what (information), to whom, when and with what (quality). In most cases, information managers are more concerned with the business aspects of information then with the IT aspects. Typically, information managers will be involved in decisions regarding investments because of their focus on business and economic issues.

From a green IT viewpoint, it is relevant that information managers can assess the "greenness" of products and processes in order to balance this aspect with other issues in investment decisions (See, for instance, the Chapter 7 written by Kalsheim and Beulen in this book). Assessing sustainability in products and processes gives the opportunity to integrate green IT knowledge in a course on information management. In the course, we ask students to perform a sustainability assessment[5] for a company that wishes to implement "cloud-based solutions". In this assessment, various elements have to be considered like the policies of the company regarding environmental issues, their commitment to climate change, energy efficiency, and renewable technology. Societal and legal factors are relevant in the case as well. The assessment will have to result in an advice to the management of the company involved.

---

[5]An overview of assessment methodologies can be found in (Sing *et al.*, 2012).

To perform this assessment successfully, students will have to combine their knowledge of IT with skills like problem solving, project management, and communication (see Section 9.3). In this way, we expect to integrate IT and non-IT knowledge, as well as hard and soft skills in the competency "understands the importance of sustainability assessments and applies these assessments in decision making".

9.5.3   *IT as enabler for sustainability*

For our third example, we will be looking towards IT as an enabler for sustainability: greening by IT.

In the built environment, it has become very important to cope with quality in relation to time pressure. Challenges for building organizations are to master the efficiency of the construction process and the communication with the customer(s). New questions as renovation and maintenance play an important role in the market and building activities increasingly take place in existing infrastructures as well.

To reduce the cost of failure and respond to the new challenges, the demand on cooperation between parties involved as well as the management of information flows becomes increasingly important. To improve the exchange of information inside and between companies, new software is currently being developed, labeled Building Information Model or Virtual Building. The aim of this software is to create a model of the building in which all relevant information for every actor involved during every phase of the building process, finds its place. With such software, it is possible to show how the resulting structure looks before the actual building has begun. Besides a much better and more timely cooperation, this gives a great advantage when communicating with the client. There are environmental and cost advantages as well: due to a more efficient development and construction process, less materials can be used in the actual construction.

In this case, IT is supporting the building process thereby creating a more efficient process with less waste. IT students may for example combine data from the BIM with other data and in this way enrich the data model of the BIM, thereby increasing their proficiency on the competency "alertness on the possibilities of IT to gain in general sustainability".

9.6   DISCUSSION

We argued that it is important to incorporate green IT issues in the curricula of IT studies as (much needed) innovation in the field usually does not come about without academic and professional research, which is a central element in the curricula of higher education. To assess which elements to incorporate in curricula, we proposed the green IT framework, which confronts the three IT-fields infrastructure, transport, and local with the product lifecycle phases design, manufacturing, use, and disposal.

We have used this framework as a starting point to formulate a set of four domain-general competencies on green IT. These competencies deal with reduction of energy consumption, materials used in IT components, disposal thereof, and greening by IT. We have emphasized the importance for every student to master these competencies in an early phase of the study, because they are the starting point for the profession-specific competencies in later years.

To establish which profession-specific competencies are necessary and which proficiency levels are within reach for these competencies, a more profound study of the domain at hand is necessary. This should be done by the experts in the educational institutes; the examples of these profession-specific competencies in various domains we have given illustrate how these competencies may be integrated in the curricula. The outcome of the learning process is not only an increased proficiency level in green IT, but it may yield organizational and societal benefits as well: it was easy for our students to propose measures that could reduce the energy usage in and around the data center with at least 20%.

So, promoting and supporting educating in green IT may be of prime importance for industry in saving energy and material cost.

REFERENCES

Coalition, E.T.: Facts and figures on e-waste and recycling. 2011, www.electronicstakeback.com (accessed January 2013).

Dochy, F. & Nickmans, G.: *Competentiegericht opleiden en toetsen: theorie en praktijk van flexibel leren.* Lemma B.V., Utrecht, The Netherlands, 2005.

EEC: European e-Competence Framework version 2.0. 2010, http://www.ecompetences.eu/ (accessed January 2013).

Gantz, J. & Reinsel, S.: Extracting value from chaos. *IDC iView*, 2011, http://www.emc.com/digital_universe (accessed January 2013).

Koomey, J.: Estimating total power consumption by servers in the US and the world. 2007, http://ccsl.iccip.net/koomey_long.pdf (accessed January 2013).

Murugesan, S.: Harnessing green it: Principles and practices. *IEEE IT Professional*, 10:1, 2008, pp. 24–33.

Plessius, H.: Duurzaamheid in ICT-onderwijs en -onderzoek. 2012, http://www.surfsites.nl/duurzaamheid/projecten/duurzamecurricula (accessed January 2013).

Pojasek, R.B.: Sustainability: the three responsibilities. *Environmental Quality Management* 19 (2010), pp. 87–94.

Sing, R., Murty, H., Gupta, S. & Dikshit, A.: An overview of sustainability assessment methodologies. *Ecological Indicators* 15:1 (2012), pp. 281–299.

Tomlinson, B.: *Greening through IT. Information technology for environmental sustainability.* MIT Press, Cambridge, MA, 2010.

Willard, M., Wiedmeyer, C., Warren Flint, R., Weedon, J., Woodward, R.I.F. & Edwards, M.: *The Sustainability Professional: 2010 Competency Survey Report.* Wiley Online Library, 2010.

CHAPTER 10

Biomimicry: Design and innovation that help reach eco-effective solutions

Saskia Muisenberg, Jaco Appelman & Dayna Baumeister

10.1 INTRODUCTION

In our search to come to useful and practical notions to stimulate attention to eco-effectiveness this chapter will connect the core concepts of biomimicry, as we understand and practice it and link the core concepts to the subject of green(-ing) ICT by giving you examples from practice. This way we come closer to giving the concept of eco-effectiveness body and plausibility. This chapter sketches the beginning of a new field of inquiry where nature provides the yardstick for progress. As such much is left to your own imagination, creativity, and innovative potential. This chapter aims to introduce and inspire, it does not provide a strategy or innovation roadmap. After an explanation of the concept of biomimicry, the different degrees of greening with biomimicry, and a very condensed overview of life's principles, the chapter ends with a presentation of cases where biomimicry was used in organizational development and manufacturing. We leave it to your imagination how you can apply it in your organization.

10.1.1 *What is it?*

Biomimicry, from the Greek bios = life; mimesis = to imitate, seeks nature's advice as a guiding framework for a transition towards sustainable living. Around the world, biomimics are consulting life and its graceful, amazing solutions to create new products, processes, and policies—new ways of living—that are well adapted to life on earth for years to come. They have learned to grow food like a prairie, adhere like a blue mussel, sequester carbon like a coral reef, create color like a peacock, and run a business like a redwood forest. Biomimicry offers a new way of looking based on what we can learn from our nature, not on what we can extract from it. This shift from learning about nature to learning from nature requires a new method of inquiry and a new set of lenses.

So, given its depth and breadth, how does one categorize biomimicry? Some say it is a design discipline, others a branch of science, a problem-solving method, a sustainability ethos, a move-ment, a way to (re)connect with nature, a new way of viewing and valuing biodiversity. It is all of those. And that is why so many people are drawn to it.

As for a more "formal" definition: "Biomimicry is learning from and then emulating natural forms, processes, and ecosystems to create more sustainable designs." For instance: "It's studying a leaf to invent a better solar cell or a coral reef to make a resilient company. The core idea is that nature has already solved many of the problems we are grappling with: energy, information storage and exchange, climate control, benign chemistry, transportation, and more. Mimicking these earth-savvy designs can help humans leapfrog to technologies that sip energy, shave material use, reject toxins, and work as a system to create conditions conducive to life."[1] A focus on biomimicry can give any company a competitive advantage, simultaneously broaden its social acceptance and reduce its footprint.

---

[1] Biomimicry Primer J.M. Benyus, http://biomimicry.net/about/biomimicry/a-biomimicry-primer/

10.2   BIOMIMICRY IS THE CONSCIOUS EMULATION OF LIFE'S GENIUS

The word "conscious" refers to intent—it is not enough to design something without nature's help and then in retrospect say, "This reminds me of something in the natural world." That's called convergent evolution, but it is not biomimicry. Biomimicry implies conscious forethought, an active seeking of nature's advice before something is designed or researched.

The nature of the consultation is also important. Seeking nature's blueprints and recipes is only part of the process; the intent should be to create products and processes and policies that fit seamlessly within the larger natural system that embody Life's Principles. This ensures that you design products, services, or systems that really mimic life and will therefore:

- use less resources,
- less energy, and
- provide better value over a longer period of time.

That is the promise when you start mimicking nature, because life creates the conditions for more life: the bedrock of most of our wealth.

The word "emulation" is more nuanced than mere copying or slavish imitation. Biomimics may study a spider to learn about sensing, fiber manufacture, adhesion, or tensegrity, but we are not actually trying to recreate the spider. What we are trying to emulate are the design principles and living lessons of the spider.

"Life's genius," a term rendered more controversial in the age of intelligent design, is also carefully chosen. It refers to the fact that biobased solutions are more than simply clever because they have evolved in response to earth's mandates. Life's true genius is in how its technologies contribute to the continuation of not just one life but all life on earth. Three levels need to be involved to come up with eco-effective designs, each level helps create organizations, technologies, products and systems that support and life.

10.3   THREE LEVELS OF BIOMIMICRY

The first level of biomimicry is the mimicking of *natural form*. For instance, you may mimic the hooks and barbules of an owl's feather to create a fabric that opens anywhere along its surface. Or you can imitate the frayed edges that grant the owl its silent flight. Copying feather design is just the beginning, because it may or may not yield something sustainable.

Deeper biomimicry adds a second level, which is the mimicking of *natural process*, or how it is made. The owl feather self-assembles at body temperature without toxins or high pressures, by way of nature's chemistry. The unfurling field of green chemistry attempts to mimic these benign recipes. An example is a material called Shrilk made from the exoskeletons of shrimps, it is as strong as aluminium and only half the weight, it biodegradable and – compatible and micro moldable.

At the third level is the mimicking of *natural ecosystems*. The owl feather is gracefully nested— it is part of an owl that is part of a forest that is part of a biome that is part of a sustaining biosphere. In the same way, our owl-inspired fabric must be part of a larger economy that works to restore rather than deplete. If you make a bio-inspired fabric using green chemistry, but you have underpaid workers weaving it in a sweatshop, loading it onto pollution-spewing trucks, and shipping it long distances, you have missed the point. To mimic a natural system, you must ask how each product fits in—is it necessary, is it beautiful, is it part of a nourishing food web of industries, and can it be transported, sold, and reabsorbed in ways that foster a forest-like economy?

If we can bioimic at all three levels—natural form, natural process, and natural system—we will begin to do what all well-adapted organisms have learned to do. Such organisms are eco-effective and create conditions conducive to life.

Creating conditions conducive to life is not optional; it is a rite of passage for any organism. If we as humans want to keep coming home to this place, we will need to learn from our predecessors how to filter air, clean water, build soil, how to keep the habitat lush and livable. It is what good neighbors do. How to put this to good effect and how to get to strategy?

10.4   LIFE'S PRINCIPLES

Life has survived 3.85 billion years of trial and error, testing, and rigorous selection. Only 0.01% of all species that have ever lived on earth survive today. Nature has some pretty high quality control standards!

This tells us there are some very powerful strategies at work for survival. Strategies that are embedded in the millions of species that live on earth today. The species thriving today are the success stories, they are the ones that managed to survive and thrive in the current context.

It may seem that every organism has its own unique strategy for surviving and even thriving in its niche. But upon closer look patterns start to emerge. Many organisms have similarities; they may have a comparable shape, store energy using the same mechanism, or have a similar response to sudden opportunities or same chemical reaction. In biomimicry, these patters are called deep principles. Not every organism employs magnetic cues to navigate, but many organisms, both in the air, on land, and in the water do use this to find their way home. Because of its frequent appearance in nature, using magnetic cues is considered a deep principle. Some principles, such as being locally attuned and responsive, are even more common across species than the deep principles and are found uniformly across almost all organisms.[2]

Janine Benyus, Dayna Baumeister, and a core group of biologists and scientists have studied, identified, compiled, and distilled scientific research to create such a set of fundamental principles named "Life's Principles". Biomimics use these Life's Principles to both drive and evaluate the sustainability and appropriateness of our designs. Life's Principles aim to represent nature's strategies that lead to success, and provide answers how life develops and grows on earth for 3.85 billion years.

The suite of principles is a result of and subject to the operating conditions of planet earth. Very briefly the conditions are as follows:

- earth is in a state of dynamic non-equilibrium,
- subject to limits and boundaries,
- it is sun, gravity, and water-based, and
- life is subject to cyclic processes.

In other words, our planet is constantly in a state of motion, always shifting in response to changing conditions and evolving so that life can endure and regenerate. There is a limited supply of water, elements, nutrients, and atmosphere that comprises our life support system.

All organisms use the sun (direct or indirect) as their main source of energy; gravity is a typical characteristic of our planet and life responds to gravity when flying, navigating, balancing, and moving. And lastly, all chemistry on Earth happens in water. Water is that important. Seventy percent of the surface of our planet is covered by water. And our cells are 70% of water as well! Water is nature's solvent and the medium in which chemical reactions occur.

Dynamic, non-equilibrium, cyclic processes involve change, predictable change. Change of seasons, day and night, the tides are all examples of cycles found on earth. This predictable change is an important operating condition in which life must thrive. And predictable change is what organizations are looking for. Every organization tries to find a balance between room to explore and innovate and the mandate to exploit to remain profitable through well-established

---

[2]Biomimicry Resource Handbook © http://biomimicry.net/educating/professional-training/resource-handbook/ also see (Baumeister, 2013).

products and processes. Examples of cycles in business are for instance: changes in preferences of consumers, adapting to new regulations, changing resource-prices, fluctuating currencies, changes in demography, etc. To do so the IT industry already, largely unconsciously, uses some of these Life's Principles.

Life's Principles represent the overarching patterns found amongst species surviving and thriving on our planet. It is important, but hard for many people, to realize that we, humans, live and function within the same operating conditions as all other organisms. But if we accept this then Life's Principles give us important clues what organisms around us are doing that make them successful. As such, Life's Principles can be used for inspiration for product and process innovations and the development of innovative strategies to come to sustainable designs.

Life's Principles also provide an ambition: fitting in on earth goes beyond our own survival. We are part of the larger system and therefore can contribute to the health of our planet. All life is interdependent and interconnected and Life's Principles show how a seemingly isolated design is linked to larger systems and how all species link as one interconnected system.

Life's Principles are at the same time sustainable benchmarks, with them we can test whether our designs meet these principles. If we purposely set out to incorporate the Life's Principles into our designs, chances are high that these designs will fit in on earth and help companies and people survive and thrive.

When you look at energy and ICT the first thing we need to know is that almost all species replicate strategies that work, and reshuffles information during design cycles to launch new apps or games or other software releases. That is the beauty of it recognizing it leads to more idea how value can be created and sustained, what principle of life can be added to the corporate equation.

A short overview of all Life's Principles is given below: we will selectively discuss a few of them to help the reader to make sense of such principle's in a business context.

Life's Principles (also see Fig. 10.1):

1. **EVOLVE TO SURVIVE**
   - replicate strategies that work
   - integrate the unexpected
   - reshuffle information
2. **BE RESOURCE (MATERIAL AND ENERGY) EFFICIENT**
   - use multi-functional design
   - use low-energy processes
   - recycle all materials
   - fit form to function
3. **ADAPT TO CHANGING CONDITIONS**
   - maintain integrity through self-renewal
   - embody resilience through variation, redundancy, and decentralization
   - incorporate diversity
4. **INTEGRATE DEVELOPMENT WITH GROWTH**
   - combine modular and nested components
   - build from the bottom-up
   - self-organize
5. **BE LOCALLY ATTUNED AND RESPONSIVE**
   - use readily available materials and energy
   - cultivate cooperative relationships
   - leverage cyclic processes
   - use feedback loops
6. **USE LIFE-FRIENDLY CHEMISTRY**
   - build selectively with a small subset of elements
   - break down products into benign constituents
   - do chemistry in water

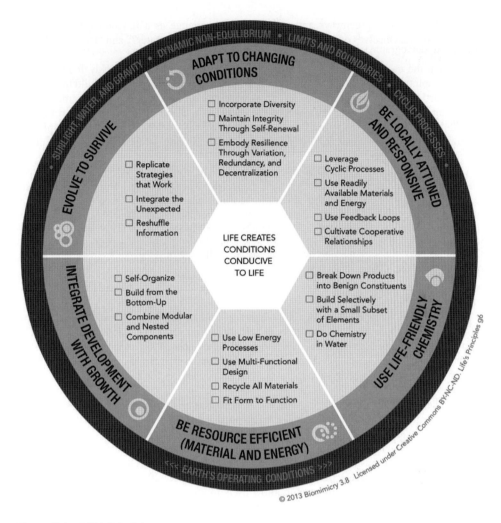

Figure 10.1. Life's Principles.

More detail on how to use Life's Principle can be found in the case description with which we end this section. It shows how IT and biomimicry can be applied to real world cases.

10.4.1 *Biomimicry case study: how can IT support sustainable communities?*

In 2010, IBM Citizenship created the Smarter Cities Challenge to help 100 cities over a three-year period to address some of the critical challenges facing cities. This program uses the company's information technology to help municipal governments create healthier, more intelligent urban environments for their residents. Using their ability to collect and analyze data, IBM is able to provide information about elements of daily city life ranging from weather and traffic to water usage and air quality. At some point IBM asked how they could use nature to understand how these overlays of information could help guide residents toward making better personal decisions for the good of the city.

The Smart Team chose to focus on water conservation. IBM's goal was to make cities more responsive and resilient. Using a biomimetic approach, encouraging more responsible water usage, could have a real impact when implemented across an entire municipal area.

Visits to two urban greenways and reframing the challenge as a functional approach, the designers asked themselves questions like "How does nature store water?" and "How does nature collect water?"

Although the designers were inspired by the natural examples, they quickly found a more actionable solution in the Life's Principles, a chart created by the Biomimicry Guild to illustrate how the earth regulates and conserves resources within its own giant ecosystem. Examining this representation of nature's complex systems, the designers realized that their solution would not come directly from an organism but from this entire system as a whole: These core principles for life could work as a metaphor for a city, inspiring and informing solutions to make a healthier and smarter environment.

After studying the Life's Principles chart, one truth became apparent: Nature has strict boundaries when it came to resources. Cities do not. Especially when it came to water, organisms had very specific ways for dealing with conservation during times of scarcity. Grass will go dormant during droughts. Birds will conserve food for the good of the group. Animals and plants have an intrinsic ability to monitor and self-regulate, whereas humans are so far removed from this cycle that they only pay attention when they experience what the designers named a "heart attack moment" where resources have been depleted and it is too late. Animals regulate based on ambient conditions—they do for themselves but also for the good of the species. So how do you make those signals apparent? How can we bring out those concepts so we are looking at those boundaries and limits?

The design team tapped their own knowledge of human behavior: "People hate the word "boundary", there's nothing that frustrates people more than strict regulations and limits". The team realized to make their solution appeal to people, they needed to create "soft boundaries of encouragement" for water conservation: Using IBM's data, Smart could design feedback loops on each layer of the city's ecosystem that would create boundaries for individuals, communities, and cities.

Using the Life's Principles as their guide, the designers began framing a three-level approach to that would provide tangible and relevant feedback loops in different layers: individual (organism), communal (species), and societal (species to species).

For the individual layer, Smart wanted a non-intrusive, yet tangible way to show residents how much water they were using. They created the Heartbeat Faucet, which provides feedback by pulsing after dispensing every 0.35 liter of water—approximately 20–30 times a minute for the typical faucet. This metered pulse would allow users to see and feel how much water they are using each time they turn on the faucet, informing everyone in the household about their behavior. The faucet would also give IBM point-of-use metering for real-time analytics.

For the communal layer, Smart focused on the concept of "leveraging interdependence", or creating cooperative behavior. For this solution they looked at the idea of a Communal Reservoir, which would encourage groups to work together. In an urban environment, Smart decided, a reservoir would translate to the water used by an entire apartment building: Water usage is typically metered for a whole building but each resident often has no idea how much or little they are contributing to the bill. This program would track your building's water usage during a fixed time period and give a community reward—from municipal tax credits to flowers for the lobby—if you come in below your target. The elevator could then serve as the "town square," with displays that show the building's reservoir level and allow the community to modify behavior on daily basis.

Finally, for the societal layer, Smart needed a way to quickly convey the city's water health to its residents and reconnect cities to the natural resources their inhabitants depend on for survival. They created the concept of MicroParks, tiny greenspaces throughout the city located next to fire hydrants retrofitted with solar powered wireless water metering systems. The MicroParks' water feed could be manipulated to reflect the cities future water supply—a lush MicroPark would communicate a healthy water supply and a withering MicroPark would let residents know that conservation is critical. These miniature green spaces can forecast city water supply health and make behavior-changing daily connections with people in a positive way.

Smart was able to use the concepts of an entire ecosystem as a metaphor for a city environment, moving from very micro examples from nature into an application like the Life's Principles that applied nature in a very macro, big-picture way. This resulted in a solution that they felt was more appropriate than the traditional biomimicry approach of inscribing specific organisms' traits upon urban infrastructure and human behaviors.

10.5   BIOMIMICRY AND GREENING BY IT

The IT industry is starting to take cues from plants and animals[3] and in this way the industry opens up a route to become eco-effective and remains profitable. For digital security, watching the way ants swarm has shown developers new ways of protecting computers from viruses. Meanwhile, Qualcomm has mimicked the iridescence of butterfly wings to make computer monitors, e-readers, and cell phone displays, amongst others, more energy efficient and usually also easier to read . Even on the tiniest scale, researchers at Xerox are looking at how to equip machinery and structures with "collaborating sensors" reminiscent of neurons—the most efficient information sharers we have—to more effectively filter and transfer information within a system.

Even though we can understand the languages of many animals, studying how members of a species use communication can help us with our technology, as scientists are showing through studying bats and their echolocation abilities. Looking at how bats use echolocation and decoding how it is done is bringing scientists closer to better ways of using sonar for everything from maneuvring robotic vehicles to finding flaws in building structures.

Many people believe that technology is going to help us save and restore the planet; it only makes sense for us to look at how the planet functions in order to design the technology that will do the saving.

Regarding biologically inspired computing, there are, for instance: genetic algorithms, neural networks, and viruses. At a genetic level nature certainly stores information and expresses this information. Further, nature does have communication systems, and sometimes very complex information systems. Clearly there are metaphors that could be used.[4]

Jefferies, a global securities and investment banking group, in an evocative description of an agent-based information retrieval system describes "this program works like a hive of bees, going out for pollen and bringing it back to the hive". In "Biomimicry for optimization, control and automation", Passino has a extensive chapter on foraging behaviors as a basis for search optimization. Clearly, there are biological metaphors that could be used in software development.

The software end of computing is about information, and there are many ways nature manages/processes information. On www.asknature.com there are currently 38 strategies organisms employ to navigate, 20 different ways to process signals, almost 100 different ways of sending signals, and almost 200 ways to sense signals/environmental cues. Some examples of bio-inspired sensors include:

- Locust-inspired collision avoidance technology (Volvo)
- Bat-inspired cane for the blind
- Jewel beetle-inspired fire detectors
- Lobster-inspired chemical detectors that can pinpoint the source of chemical pollution in water

Nature can also teach us a great deal about feedback loops. The true power of feedback loops is not to control people but to give them control. "It's like the difference between a speed trap and a speed feedback sign—one is a game of gotcha, the other is a gentle reminder of the rules of

---

[3]http://www.treehugger.com/clean-technology/top-5-areas-where-nature-inspired-innovation-works.html, Jaymi Heimbuch on Treehugger.
[4]http://computingforsustainability.com/2007/08/27/biomimicry-in-software-engineering-a-super-system-metaphor/

the road."[5] The ideal feedback loop gives us an emotional connection to a rational goal. And IT can do just that, by giving real-time, relevant information on our behaviors. An example of this is Belkin's Conserve Insight (see Text box 10.1).

Belkin's Conserve Insight is an outlet adapter that gives consumers a close read of the power used by one select appliance: Plug it into a wall socket and then plug an appliance or gadget into it and a small display shows how much energy the device is consuming, in both watts and dollars. It is a window onto how energy is actually used, but it is only a proof-of-concept prototype of the more ambitious product Zorro.

In pure feedback terms smart meters fail on at least two levels. For one, the information goes to the utility first, rather than directly to the consumer. For another, most smart meters are not very smart; they typically measure overall household consumption, not how much power is being consumed by which specific device or appliance. In other words, they are a broken feedback loop.

Belkin's device avoids these pitfalls by giving the data directly to consumers and delivering it promptly and continuously. "Real-time feedback is key to conservation", "There's a visceral impact when you see for yourself how much your toaster is costing you". Belkin predicts that home sensors will one day inform choices in all aspects of our lives. "We're consuming so many things without thinking about them—energy, plastic, paper, calories. In the future we will have an ubiquitous sensor network, a platform for real-time feedback that will enhance the comfort, security, and control of our lives".

Text box 10.1. Greening by IT with Biomimicry

And today, the promise of feedback loops could not be greater. The intransigence of human behavior has emerged as the root of most of the world's biggest challenges. Witness the rise in obesity, the persistence of smoking, the soaring number of people who have one or more chronic diseases. Consider our problems with carbon emissions, where managing personal energy consumption could be the difference between a climate under control and one beyond help. And feedback loops are not just about solving problems. They could create opportunities. Feedback loops can improve how companies motivate and empower their employees, allowing workers to monitor their own productivity and set their own schedules. They could lead to lower consumption of precious resources and more productive use of what we do consume. They could allow people to set and achieve better-defined, more ambitious goals, and curb destructive behaviors, replacing them with positive actions. Used in organizations or communities, they can help groups work together to take on more daunting challenges. In short, the feedback loop is an age-old strategy revitalized by state-of-the-art technology. As such, it is perhaps the most promising tool for behavioral change to have come along in decades. IBM case studies of biomimicry and IT. The first is related to supporting sustainable communities, the second to how nature inspires computer chip manufacturing.

The first IBM case (see Text box 10.2) shows how Life's Principles were helpful in generating and organizing multi-level change and what the roles IT can play in this respect.

IBM brings nature to computer chip manufacturing[6]

Armonk, NY—03 May 2007: IBM (NYSE: IBM) today announced the first-ever application of a breakthrough self-assembling nanotechnology to conventional chip manufacturing, borrowing a process from nature to build the next generation computer chips. The natural

---

[5]http://www.wired.com/magazine/2011/06/ff_feedbackloop/2/
[6]http://www.weforum.org/issues/global-information-technology

pattern-creating process that forms seashells, snowflakes, and enamel on teeth has been harnessed by IBM to form trillions of holes to create insulating vacuums around the miles of nanoscale wires packed next to each other inside each computer chip.

In chips running in IBM labs using the technique, the researchers have proven that the electrical signals on the chips can flow 35% faster, or the chips can consume 15% less energy compared to the most advanced chips using conventional techniques. The IBM patented self-assembly process moves a nanotechnology manufacturing method that had shown promise in laboratories into a commercial manufacturing environment for the first time, providing the equivalent of two generations of Moore's Law wiring performance improvements in a single step, using conventional manufacturing techniques.

This new form of insulation, commonly referred to as "airgaps" by scientists, is a misnomer, as the gaps are actually a vacuum, absent of air. The technique deployed by IBM causes a vacuum to form between the copper wires on a computer chip, allowing electrical signals to flow faster, while consuming less electrical power. The self-assembly process enables the nanoscale patterning required to form the gaps; this patterning is considerably smaller than current lithographic techniques can achieve.

Self assembly is a concept scientists have been studying at IBM and in labs around the world as a potential technique to create materials useful for building computer chips. The concept occurs in nature every day, it is how enamel is formed on our teeth, the process that creates seashells and is what transforms water into complex snowflakes. The major difference is, while the processes that occur in nature are all unique, IBM has been able to direct the self-assembly process to form trillions of holes that are all similar.

This new technology can be incorporated into any standard CMOS manufacturing line, without disruption or new tooling. The self assembly process was jointly invented between IBM's Almaden Research Center in San Jose, CA and the T.J. Watson Research Center in Yorktown, NY. The technique was perfected for future commercial production at the College of Nanoscale Science and Engineering of the University at Albany, within the world-class Albany NanoTech facilities, a research and development site in Albany, NY with strong ties to IBM, and at IBM's Semiconductor Research and Development Center in East Fishkill, NY.

Text box 10.2. Life's Principles and strategy

Nature can teach us how to optimize and manage the lifecycle of everything around us. We can optimize the flow of people, water, electrons, materials, and other natural resources. Many say that IT innovations have led to the human-nature separation. But we can also use IT further in our advantage. As we live in a connected world, IT provides us the means to do this monitoring, managing, and controlling of flows.

10.6 CONCLUSION

The amazing insight is that nature holds the answers to many of our problems already, as long as you are ready to go out and consciously learn from and then emulate from nature. In Chapter 1.4 of the Global Information Technology Report[7], Cesar Alierta focuses mainly on ICT as the platform, which brings about the democratization of information. But in reality, information is already everywhere around us, in nature, just waiting to be picked up and used to solve problems. Biomimicry holds great potential for innovations in the field of ICT itself (bio-inspired hard- and software) and at the same time ICT will play a large role in designs incorporating Life's Principles.

---

[7]http://www.weforum.org/issues/global-information-technology

When we are able to increase the extent to which the impact of new technologies can benefit the economy and society at large, we can create a hopeful outcome for all of us. Luckily we are not on our own. We have over 30 million elderly species waiting out there to be consulted.

REFERENCES

Baumeister, D.: *Biomimicry Resource Handbook: A Seed Bank of Best Practices*, Missoula, MT: Biomimicry 3.8. 2013.

Benyus, J.: *Biomimicry: innovation inspired by nature*. Perennial, New York, 2002.

Liu, Y. & Passino, K.: Biomimicry of social foraging bacteria for distributed optimization: models, principles, and emergent behaviors. *Journal of Optimization Theory and Applications* 115:3 (2002), pp. 603–628.

Passino, K.: *Biomimicry for optimization, control, and automation*. Springer, 2004.

CHAPTER 11

Conclusions: Exploring synergies between efficiency and effectiveness

Anwar Osseyran, Jaco Appelman & Martijn Warnier

There is an urgent need for effective measures to stop the explosive growth of global energy consumption and related environmental impact. Information and Communication Technologies (ICT) play since many decades a central role in increasing efficiency and improving effectiveness of energy usage in many vital sectors of the economy. This has however led to an explosive growth of the energy used by the ICT sector itself and the call for decreasing the carbon emissions of ICT. On the other hand, improving efficiency of energy use has also led to an absolute increase of consumption and of carbon emissions (Jevons paradox). In this book, we present a holistic approach for Green IT: working on two parallel action lines to green the IT sector itself (Greening IT) and deploying ICT to green the other sectors (Greening by IT).

How to create synergy?
As a transition towards sustainability is imperative, different approaches, methods and tools to support this transition were adopted and this has been acknowledged in this book. Despite the common goal, this diversity threatens to slow the process of transition rather than accelerating it. On the other hand, adopting and implementing a combination of approaches could lead to the exploitation of complementarities and providing synergy rather than causing deadlocks. In order to compare the different sustainability approaches we have adopted a set of selection criteria and a common frame of reference based on literature and on the results of an expert meeting. The approaches we selected had to be scientific based, generic and not limited to a specific discipline and leading to a clear process of transition towards sustainability. Seven approaches complied with our criteria for selection: The Natural Step (TNS), Cradle-to-Cradle (CtoC), Ecological Footprint (EF), Biomimicry (B), Industrial Ecology (IE), Natural Capitalism (NC) and Corporate Social Responsibility (CSR). Our scope in this book was however restricted to the first three (TNS, CtoC and EF) due to the local focus of the sustainability community in The Netherlands.

The most important element of a sustainability approach as identified by the expert meeting described in the book includes inspiring vision, community feeling and cooperation, awareness, clear science based criteria for sustainability and systemic change. The study presented in this book shows that the combination of the three selected approaches leads to complementarity: TNS offers the needed strategic framework and process required to implement the vision presented by CtoC, and EF delivers tools for measuring baseline conditions and progress. The combination of the three approaches was also mapped into the ICT sector itself: EF is leading to quantitative measures of the ICT ecological footprint, ICT is helping modern businesses to adopt efficient, cyclic processes better fitting within TNS and CtoC is the focal point when it comes to disposal and re-use of hardware and software systems. The analysis presented in this book shows that other combinations are also possible and probably very useful. The selected approaches still need to pay some attention to behavioral and spiritual elements that cannot be ignored as those are needed to counterpart thoughtless consumption trends and powerful vested interests.

A strategic view on Green IT
Regardless of whether we use one approach or a combination of approaches, the transition towards sustainability must deal with greening the ICT sector itself, as with deployment of ICT to green other sectors. Greening IT requires us to deal with the datacenter where as a result of virtualization,

most of the energy is being consumed. The clouds will make computing and data access ubiquitous and open the way to increase the utilization factor of datacenters, therefore increasing their energy efficiency. Other important measures in greening IT are greening the software and data usage as those have a direct magnifying impact on application footprint and over-sizing of energy-consuming hardware.

Greening by ICT has already been done for many decades and ICT should be positioned at the center of the rapidly growing green economy. ICT helps making the electricity grid smart, with the goal of optimization of electricity consumption, decentralization of energy production and deployment of renewable energy sources. ICT helps greening the transport sector by optimizing logistics, improving filling rates, reducing congestions and minimizing unnecessary transport of people and materials. Smart buildings and smart cities use ICT for getting better insight in energy consumption and enabling rethinking of building designs and urban plan making to provide macro-solutions rather than seeking sub-optimizations. Greening by IT helps also to green the industry and stop the overshoot of raw material consumption, by optimizing supply chains and raw material (re-)use and leading to customer participation as a prosumer of the industry.

Big Data will help us better understand people behavior and assist in improving the adoption and deployment of sustainability measures. Dematerialization and avoiding a rebound in energy consumption as a result of improving energy efficiency, require therefore a holistic approach in green IT implementations that goes beyond the first order impact of ICT.

Higher order Impacts of Green IT

While the focus of many studies is on the first order effects of ICT on sustainability, this book dealt also with the higher order impacts of ICT on sustainable development that are frequently associated with the short life cycles of this relatively young sector. Beside the known second order rebound effect (or Devon Paradox) that increased efficiency will lead to a strong increase in demand and hence a net increase of consumption, there are other unintentional or unpredicted secondary effects of ICT. On the other hand, it seems that a rebound effect may also follow in some cases an evolutionary path leading after all, to a net decrease in consumption due to the use of ICT. Many examples of secondary effects of ICT have been given including changes in Job profiles of professions; changes in markets and company profiles; impact of dematerialization on material consumption, IPR earnings and proliferation and implicit promotion of standards and tools; decrease of diversity in product design, communication languages and push for globalization; and higher order impact of all e-applications of ICT.

Most of those higher order effects are related to dematerialization and ease of exchange of data, information and expertise offered by ICT. Expected sustainability improvements should be thoroughly analyzed and monitored in order to avoid failure, rebound of unwanted higher order effects. This requires sustainability training and expertise in most of the ICT application areas. A balanced assessment of those higher order sustainability effects is frequently neglected or intentionally ignored due to vested interests or focus on short terms earnings. We have shown that those higher order effects can be both positive and negative to sustainability but when we identify them in the early stages we would be able to act timely. The indirect sustainability impacts of ICT are not negligible and must be dealt with adequately.

Multi-order impact of Standardization

Standardization activities have also first and second order effects on sustainability. Standards are mostly motivated by economic purposes and are needed to enable healthy growth of market volumes of mostly new technologies like ICT. The side or second order effect of standards in ICT is that they help reducing variety, forcing efficiency through open competition and lock-in prevention, and building critical mass. Standards help also innovation by opening participation to new players and reduce unnecessary and harmful diversity making markets transparent. All leading to positive implications for the environment as excessive diversity and incompatibilities of technologies in ICT generate e-waste, discourage re-use and make recycling economically unviable. In this book, we have therefore presented an economic-environmental framework for analyzing sustainability effects of compatibility standards. The framework has been applied to the mobile phone chargers market as a test case.

Most important ICT standards that are sustainability-related are the management standards of the ISO 14000 and ISO 26000 series, and the standards for collecting data on, for example, climate change. Most focus is put on the use of standards as a means to improve eco-efficiency. Eco-effectiveness is expected to be a side-effect of compatibility standards, as those standards will push improving sustainability in the product design. Eco-design is as important as the energy and carbon emissions of ICT products are for a large part determined in the design phase. The positive environmental side effects of compatibility standards coincide with three design principles of the CtoC approach described before: elimination of the concept of waste; no downgrading of material should occur; and the product life cycle is material- and energy-efficient. The standardization of plug and socket of mobile phone chargers has led to the avoidance of vendor lock-in, facilitated growth of re-use markets for chargers and mobile phones, provided incentives for recycling and made it easier for consumers to influence the market and favorize eco-designs. Further research is recommended in order to examine whether compatibility standardization can be regarded as a form of eco-design at sector level. More case studies and quantitative data are also needed to determine whether the environmental effects we have identified are correct, complete, and to calculate their environmental impact and examine possible rebound effects. We also recommend a systematic analysis of circumstances under which standards should be imposed because of the risks of market failure of self-regulation.

The magic of Micro-Agreements
A second order effect, not of ICT but of local sustainable energy production, is that those reveal to be difficult to manage without ICT and automation. Despite the growth of investments in renewable energy production, those investments are still mostly underutilized and the efficiency of deployment is stuck around 10% due to the complexity of matching intermittent generation of sustainable energy with demand. Energy storage technology is still in its infancy and the only solution we propose in this book is a dynamic matching of supply and demand in a smart grid offering bi-directional flow of energy and information and supply-demand management. Economic incentives offer the tool to manage supply as customers can postpone consumption when energy is scarce and thus expensive. Since sustainable energy sources (like wind and sun) are typically intermittent, the decision-making window is short and the process is complex and dynamic. Pilots exploring Real Time Pricing schemes managed to reduce energy load by up to 33%. But users failed to respond to hourly changes in prices. Other problems were the lack of transparency of prices or user-friendliness of the system. The process must therefore be automated and software agents can help in this, monitoring and responding to real-time information about (sustainable) energy availability and market price, quickly and efficiently. Negotiations can then be automated and micro-agreements between energy suppliers and consumers can automatically be closed. This new system is expected to reduce wasted energy overcapacity, lower prices and increase utilization of green resources.

Transparency and trust are crucial for success of this smart grid system. The software agents should therefore represent all the players on the energy market including consumers, private and professional energy producers of all kinds, mediators and brokers, and energy operators and distributors. The smart grid system should be under public supervision and must comply with certification requirements to ensure trust and transparency. The software agents are expected to have access to the grid information including usage data, user behavior, preferences and decision rules of the parties they represent. The smart grid disposes also over the necessary ICT-means to store and process all the auctions and transactions. The smart grid system we propose offers benefits to all players involved, the consumer can favorize sustainable energy production, become a producer himself, strengthen his negotiation position and lower the total bill of his consumption. The producer can reduce overcapacity, reduce capital expenditure, operation expenses and minimize waste. The community benefits as well through a macro-optimization of capital expenditure, energy use, sustainable energy deployment and carbon footprint. We therefore recommend to conduct future research work on how to simulate this micro-agreement energy marketplace using software agents, negotiation protocols and real-world usage ICT technologies like artificial intelligence and big data.

"That which is measured, improves"

Measuring environmental efficiency of IT and setting strategies for green IT as part of corporate responsibility within an organization require a holistic approach. Greening IT can be done step by step and a framework is needed to measure progress and help select the most effective approach. The framework should cover economic, environmental and social requirements and incorporate assessment criteria with related performance indicators to support quantifying, and therewith justifying the Green IT decision making. Performance Indicators (PIs) should be repeatedly measurable in time, target-oriented, comprehensible, comparable, compatible, and balanced. PIs must also reflect environmental performance adequately and display problems next to benefits. In this book we presented such a framework. The requirements for our green hardware IT Infrastructure (GHITI) framework have been elicited from literature and through semi-structured interviews and a brainstorm session with relevant stakeholders and experts. The framework covers three stages: procurement, use and disposal of IT, and incorporates various performance indicators of resource usage (energy, water and raw materials), GHG emissions, waste electronic and electric equipment and costs related to IT hardware supporting business applications. A case study has been conducted as part of evaluating the GHITI framework and the evaluation results were presented. Despite the narrow scope of developed framework and limitations of case study, the evaluation shows that it is important and necessary to develop a framework tool that is flexible enough to adapt to changing requirements and that performance indicators could be measured and aggregated into one value stating the relative greenness of the IT infrastructure, facilitating Green IT decision making within organizations.

The presented framework enabled identification of main Green IT drivers, harvesting low hanging fruit making decision-making trade-offs visible. The highest value is obtained when the framework is made part of a continuous management process for Green IT improvement over time. A cyclic development was therefore recommended as making progress using an incomplete framework was found to be much more effective than to focus on a full measurement of the entire IT infrastructure using an all-encompassing framework. Over time additional performance indicators can always be added to the framework in use. We also formulated guidelines to adhere to, when using the framework. Those include keeping it up-to-date; tailoring information to its target audience; exploiting most available information; making sure the users do possess needed basic Green IT knowledge; continuously communicating and extending the user base and making the framework an integral part of the organization's environmental performance assessment. The GHITI framework developed and presented in this book is far from complete. We therefore recommend that additional design cycles are implemented to further verify, validate and detail the framework design and to perform additional case studies. The extended framework must also incorporate a social dimension in order to improve the balance of the three sustainability dimensions: economic, environmental and social.

Micro-trainings to support Innovation

The road towards sustainability requires continuous innovation. Green IT innovations aim for more efficient and effective need-fulfillment and diminished resource consumption and pollution, both of products and of the production of these products. To achieve Green IT innovations, research has to be conducted and newly built expertise needs to be shared. Formal disciplinary training programs are less suitable for this purpose. We introduced in this book the micro-training method that showed to be a suitable method for supporting Green IT innovations within organizations. Innovation requires creativity, taking risks and a minimum of active participation of the various layers within an organization. As it is mostly driven by the 'informal' organization and not top-down, learning should not be limited to the 'formal innovators' in an organization. The process of getting support from top to bottom in the organization is one of the hardest tasks in the whole endeavor. Continued and targeted trainings at all layers are therefore influential in the success of innovation, not only for learning new things but especially for un-learning legacy methods and reflecting on new outcomes. Incremental changes are important making low-hanging fruit easy to grasp but will be insufficient in the long run. Radical changes with a longer payback time are more difficult but are needed to realize an eco-leap. The challenge for sustainable innovations is hence to seek a long term perspective that allows reduced resource consumption and reduced pollution, helping all stakeholders understand their work in a wider context.

Learning on the other hand is promoted when learners are engaged in solving real-world problems, their knowledge is activated as a foundation for new knowledge, and new knowledge is demonstrated, applied and integrated. Effective trainings or instructions in the organization should therefore make sense to the participants (appropriateness), be of interest to them (motivation) and be easy to remember and apply (effectiveness). Since it is often not clear what has to be learnt in innovation processes, trainings must support flexibility and sharing both tacit and formal knowledge. We advocate in this book the deployment of short training sessions that are facilitated by non-professional trainers, with topics defined by the innovators involved, taking place on or near the workplace, organized at the appropriate moments and usable for various target groups. The aim is not merely focusing on individual learning, but to support change. The short duration of the sessions offers flexibility in planning, which links to the principle of learning at the point of need, or so called 'teachable moments'. An example of a 15 minutes micro-training session would start with addressing an actual Green IT fact or raising a sustainability question, followed by an exercise or demonstration, and discussion including feedback. It ends with directions for further development and a brief preview of the next sessions. The sessions are decentralized and have an ad-hoc nature. Two cases were presented showing how help organizations address sustainability topics at the point of need, widening the employees' perspective, and making them see the advantages of alternative approaches, not just the risks. However, micro-trainings will never replace formal training courses that provide employees with the deep foundations of their knowledge but will help generate support for sustainable innovations. Green IT must therefore become part of every IT curriculum.

Green IT in Education
In order to assess what is relevant from an educational point of view, we introduced a Green IT Framework that reflects the major sustainability competencies that should be embedded in an IT curriculum and how these relate to the IT lifecycle. In this framework, IT components consisting of any combination of hardware, software, data and knowledge, are categorized across three fields: shared infrastructure (datacenter or cloud-based); end-user local IT like desktops, laptops, thin clients, printers but also end-user applications, apps and locally stored data; and so-called transport IT with all components used for the exchange of data between various nodes of a network, like routers, switches and cabling, but also drivers and protocols like IPv6 as well. We divided the lifecycle of these components in four phases: design, manufacturing, use and disposal and assessed sustainability aspects of the components with regard to all relevant aspects in terms of energy and material usage, exploring its relevance for inclusion in an IT curriculum. We have also adopted a scale of 0 to 4 to characterize the level of proficiency required: awareness (0); basic (1); advanced (2); professional (3) and expert (4).

Bringing Green IT elements into higher education IT curricula, we decided to discern between profession-specific competencies and more domain-general competencies. Domain-general competencies are important for every IT professional but their proficiency level is generally awareness sometimes basic on the defined scale. Profession-specific competencies on the other hand are dependent on the line of work and may require a proficiency level of advanced to professional. Students should hence at least comply with the following Green IT competencies: awareness of the energy consumption of IT in daily use by hardware, software and applications; ability to measure energy consumption of local IT components and where possible, to reduce it or choose low-energy alternatives; become familiar with the constituents of IT components and favorize the most sustainable components to reduce e-waste and be aware of the possibilities of IT to green other sectors offering Green IT solutions. Sustainability assignments to undergraduate students in a virtual world were developed in order to integrate the green IT competencies described above, in existing courses. The assignments led to measures that could deliver a reduction in energy consumption of at least 20%, and helped the students reach the desired sustainability proficiency level as well. The framework revealed to be a good starting point for developing a set of four general-domain competencies on Green IT: reduction of energy consumption; materials used in IT components; disposal thereof and greening by IT. It is important for every student to master these competencies in an early phase of the study, and it is necessary that profession-specific competencies and their proficiency levels are identified by educational institutes and integrated in the curricula. So, promoting and supporting Green IT in education will not only yield an increased proficiency level in sustainability, but will also deliver organizational and societal benefits as well, such as competitive skills and important savings in energy and material costs.

Finally, nature is our best teacher
One of the most promising sustainability educational approaches is learning from nature itself. In this book, we propose to use biomimicry to seek nature's advice as a guiding framework for our Green IT approach. Biomimicry is "imitating life", learning from nature and then emulating natural forms, processes, and ecosystems to create more sustainable designs. The goal is to have only products, processes and policies that fit seamlessly within the larger natural system and that embody "Life's Principles" as studied, identified, compiled and distilled by a core group of biologists and scientists. Using Biomimicry ensures that products, services or systems will mimic life and therefore use fewer resources, less energy and provide better value over a longer period of time. "Biomimicking" should then be done at three levels: individual (organism), communal (species), and societal (ecosystem). Mimicking individual organisms is more and more common in product development, but the challenge is to elevate the analogy towards the process and ecosystem. Mimicking similarities between organisms, so-called deep principles, help overarching patters found amongst species surviving and thriving within the eco-system on our planet.

The IT industry is already, largely unconsciously, using some of these life's principles and our goal is to extend Green IT towards their adoption and conscience use. Life's Principles help inspiring product and process innovations and the development of innovative strategies to come to sustainable designs. Life's Principles also provide an ambition to fit within our ecosystem and provide us with reliable sustainability benchmarks. Applying Life's Principles in ICT would imply that we must replicate strategies that work, and reshuffles information during design cycles before launching for instance new apps or games or other software releases. Principles like natural evolution, resource efficiency, adaptation to change, integration of development with growth, being responsive to local ecosystems and using life-friendly technology can help Greening IT and using IT to green other sectors. There is a lot of information storage, exchange and processing in nature and the analogy can be easily drawn towards Green IT. Biomimicry teaches us that nature holds the answers to many of our sustainability problems and how to integrate those within our Green IT approaches.

Subject index

Sustainable Energy Developments

Series Editor: Jochen Bundschuh

ISSN: 2164-0645

Publisher: CRC Press/Balkema, Taylor & Francis Group

1. Global Cooling – Strategies for Climate Protection
   Hans-Josef Fell
   2012
   ISBN: 978-0-415-62077-2 (Hbk)
   ISBN: 978-0-415-62853-2 (Pb)

2. Renewable Energy Applications for Freshwater Production
   Editors: Jochen Bundschuh & Jan Hoinkis
   2012
   ISBN: 978-0-415-62089-5 (Hbk)

3. Biomass as Energy Source: Resources, Systems and Applications
   Editor: Erik Dahlquist
   2013
   ISBN: 978-0-415-62087-1 (Hbk)

4. Technologies for Converting Biomass to Useful Energy –
   Combustion, gasification, pyrolysis, torrefaction and fermentation
   Editor: Erik Dahlquist
   2013
   ISBN: 978-0-415-62088-8 (Hbk)

5. Green ICT & Energy – From smart to wise strategies
   Editors: Jaco Appelman, Anwar Osseyran & Martijn Warnier
   2013
   ISBN: 978-0-415-62096-3

6. Sustainable Energy Policies for Europe – Towards 100% Renewable Energy
   Rainer Hinrichs-Rahlwes
   2013
   ISBN: 978-0-415-62099-4 (Hbk)